Integrated Product Design and Manufacturing Using Geometric Dimensioning and Tolerancing

CRC Press
Taylor & Francis Group
Boca Raton London New York

CRC Press is an imprint of the
Taylor & Francis Group, an **informa** business

MANUFACTURING ENGINEERING AND MATERIALS PROCESSING
A Series of Reference Books and Textbooks

EDITOR

Ioan Marinescu
University of Toledo
Toledo, Ohio

FOUNDING EDITOR

Geoffrey Boothroyd
Boothroyd Dewhurst, Inc.
Wakefield, Rhode Island

Additional Volumes in Preparation

Integrated Product Design and Manufacturing Using Geometric Dimensioning and Tolerancing

Robert G. Campbell
William Rainey Harper College
Palatine, Illinois

Edward S. Roth
Productivity Services
Albuquerque, New Mexico

CRC Press
Taylor & Francis Group
Boca Raton London New York

CRC Press is an imprint of the
Taylor & Francis Group, an **informa** business

CRC Press
Taylor & Francis Group
6000 Broken Sound Parkway NW, Suite 300
Boca Raton, FL 33487-2742

First issued in paperback 2019

ISBN-13: 978-0-8247-8890-2 (hbk)
ISBN-13: 978-0-367-39583-4 (pbk)

Visit the Taylor & Francis Web site at
http://www.taylorandfrancis.com

and the CRC Press Web site at
http://www.crcpress.com

To my father, who believed as Ed Roth does that manufacturing is both an interesting and honorable way to spend one's life. And to my mother, who worked all her life and made my engineering education possible.

RGC

Foreword

In 1983, while working at Ford Motor Company, my colleagues and I had the pleasure of meeting Ed Roth at one of his seminars. He opened our eyes to the linkage needed for communication to take place between Product and Manufacturing Operations. This was done via an excellent geometric dimensioning and tolerancing (GD&T) standard, ANSI Y14.5M-1982, and in a manner we had never seen before—not just the use of GD&T, which we had already been using for 25 years, but how to effectively apply it. He also showed us the importance of GD&T's being interpreted correctly by manufacturing. The efficient application of GD&T to the product development process enhanced productivity tremendously and yielded long-term benefits. The ANSI Y14.5M standard was the tool needed to define and communicate a product definition that everyone could read and utilize. As a result, a massive training program was initiated at Ford to effectively apply GD&T throughout operations worldwide. I had the pleasure of working closely with Ed Roth for many years and gleaned some of his expertise, which made my engineering life easier and less stressful.

As a product design and manufacturing engineer, I often struggled with communicating product definition, design intent, manufacturing, and quality requirements. The drafting room had its protocols and standards for definition. Design intent regarding functional requirements was often overlooked. Manufacturing utilized their expertise, past practices, and available facilities to produce the part. After the part was made, Quality verified that all drawing dimensions had been achieved. Basically, each activity group operated independently with only the drawing specifications as the common denominator.

Unfortunately, GD&T was not applied effectively. All too often, I received a finished part that failed to meet functional requirements. If a part was "to print," but still failed to assemble with its mating part, the drawing was modified verbally to include specific functional criteria. Sometimes inspection gage requirements were defined. The knee-jerk reaction of many engineers was to tighten tolerances. If this failed, a note was added stating, "Must Meet Functional Requirements." The part supplier was held accountable for characteristics that were not clearly defined.

This book documents methodologies that can revolutionize engineering and manufacturing, provided no short cuts are taken. GD&T based on a national standard is the "language" for communicating product specifications, but *teamwork* among the various functional activities involved with the product development process must be achieved for effective "application" to achieve optimal benefits. This book provides various techniques to accomplish this feat. It is not a quick fix, but a permanent fix for longstanding problems.

Several elements were essential to a viable program in GD&T at a large corporation.

> *GD&T Standard.* We had to adopt a common standard for dimensioning and tolerancing and apply it throughout Engineering and Manufacturing Operations worldwide. This we did by endorsing ANSI Y14.5M-1982 totally and subsequently adopting the 1994 update as the corporate standard for GD&T.
>
> *Corporate Expertise.* A task force, which eventually became a permanent group, was established to gain GD&T expertise, develop training programs, and establish a company focal point for consulting and assistance to line operations worldwide.
>
> *Local Expertise.* Local operations needed GD&T experts to provide ongoing day-to-day guidance. Some of the local experts subsequently became instructors in GD&T for their respective areas.
>
> *Massive Training Program.* A very energetic training program was launched and still continues to upgrade everyone's understanding, application, and interpretation of GD&T. This included involving personnel from Product, Design Services, Quality, Suppliers, Manufacturing Processing, and Tooling/Gage Design in joint training sessions. This was a Ford first at that time (1983).

I previously referred to GD&T as a language. Consider ANSI Y14.5M-1982 the dictionary. You can't expect to hand someone a new Webster's or Oxford English dictionary and expect to create a best-selling novel without training.

You must be able to master the language to create a novel, and even then not everyone will have a best-seller. The same holds true for GD&T training and skill levels. The following four levels of expertise should be considered to ensure correct application of GD&T.

Level One. Understand the "vocabulary." All the symbols and the definitions must be taught. These are the basics.

Level Two. Usage of the symbols to form "grammatically correct sentence structure." Feature control frames must follow stringent rules to ensure correct interpretations. Datum reference frames must be depicted and used correctly. Typically this is referred to as correct syntax.

Level Three. The part definition must depict a "story," which is design intent. Mentally, the callouts must develop a cohesive plan for what will happen "pictorially" when that part is inspected for acceptance. Will the callout give the desired requirement for meeting design intent? This is a very tough question. Metrology and measurement strategy are also included for the design to be considered feasible.

Level Four. Manufacturing personnel must be able to "read" the drawing callouts and clearly understand the requirements conveyed. They must also be able to correctly apply that interpretation to select the required metrology equipment. This is also a very difficult area which requires a deep understanding of GD&T definitions. Subtle differences in data measurement methods can totally change the result. For example, do they understand the difference between measuring an axis for parallelism versus position on a CMM machine?

In implementing this training the common question that must be answered is "Who in your activity requires this knowledge?".

All activity teams must cooperate and communicate to achieve success. It is truly a team effort. I applaud the efforts and expertise that went into creating this book in which Bob Campbell and Ed Roth combine their vast experience. If you read and apply its content carefully, you will enhance your product and productivity capabilities exponentially.

Ed Boyer (Retired)
Manufacturing Engineer GD&T Specialist
Corporate Product and Manufacturing
Ford Motor Company
Bellaire, Michigan

Preface

While manufacturing firms search for a competitive edge, they frequently overlook their most valuable asset, the product development process. When rushing to apply the latest management techniques, it is easy to forget that the product must eventually be defined with real numbers. The product definition and manufacturing plans that are the results of the development process are tangible assets as important to success as production facilities and finances.

This book presents an alternative to the traditional means of developing and delivering a manufactured product. It presents an integrated product development process that emphasizes the product definition using geometric dimensioning and tolerancing (GD&T). The overarching theme of the book is the preponderant influence the product definition has on the downstream phases of the product cycle. The majority of the cost structure is locked in by the product design since the manufacturing and inspection methods are specified—to a greater degree than recognized—by the endproduct specification.

Current product design techniques do not adequately recognize the constraints the product definition imposes on manufacturing and inspection. As discussed in the first chapter, the product design is a script used throughout the development process. An inadequate script will prevent the most elaborate production plans from succeeding. For a manufactured product, the definition, including specification of the manufacturing and inspection processes, is the script.

If the firm is to succeed in delivering an economically viable product, both product and process must be fully defined prior to implementing production.

While many products have attributes that are intangible, most products have an ideal geometric form that must be translated into physical reality. The methods presented in this book are necessary to create the functional (i.e., physical) elements of part geometry and lead to efficient and economical production. Furthermore, these methods provide this information in a timely manner, enhancing the likelihood of success. The methodologies described in this book accomplish this by considering the variation encountered in the production process. Doing this early in the development cycle allows prediction of the effects that variation will have on product function and manufacturing efficiency. Designs can then be altered to remove sensitivity to specific sources of variation. At a minimum, knowledge of this variation will quantify elements of risk in the design so that intelligent business decisions can be made.

The book emphasizes a product definition that provides detailed designs—the numbers—allowing technical solutions based on objective methods rather than on visceral responses. ''Getting to the numbers'' requires both a team structure and a common language (GD&T) with which the team members may communicate. This language is necessary from an engineering standpoint to provide the dimensions used in achieving part function. It also allows the budgeting of variation among the various components comprising the assembly. The language and the methodology tie together the design and manufacturing processes, formally eliminating the artificial barriers that segregated these functional areas in the past.

The key to understanding the book's structured design techniques is the knowledge that all the information that these techniques generate will eventually be identified and documented for a successful product. The greatest economic benefits will be achieved if the information is identified and documented early in the design cycle—both accomplished using GD&T. The premise of the book is that this happens only through the use of the structured design methodologies and GD&T. If the information is generated by trial and error during the latter stages of the cycle, the price is high in terms of lost market opportunities and avoidable expenses.

Because the product definition affects the entire product development cycle, the book is directed toward design, manufacturing, and quality assurance personnel involved with product and process design. It provides a set of structured methodologies that can be used in developing specific product, manufacturing, and inspection designs that take the product from the designer's virtual world of perfection and into the physical world of variation. The common thread that runs throughout the book is the use of geometric tolerancing based on ASME Y14.5M and equivalent ISO standards.

An additional audience for the book is to be found in the engineering schools of universities. The current university environment usually treats the engineering graphics course as a dead-end vehicle; concepts initially developed in the first graphics course are not integrated into subsequent courses. This book provides an instructional framework with which to expand graphics concepts into advanced undergraduate or graduate courses. The book's emphasis on product definition with the overarching focus on variation enhances a student's ability to create an interpretable and producible product definition. Once mastered, this material ultimately makes the student more valuable to an employer.

The book is composed of five major parts. For individuals with adequate knowledge of both GD&T and manufacturing, many of the chapters can be read separately. Chapters 1 through 3 set the stage and provide the foundation concepts that support the integrated product development system. Some of this information is in other sources but is not placed within the context of a product development system or described in sufficiently concise form to show the connection between part geometry and the development cycle.

Once the stage is set, Chapters 4 through 6 provide the structured design techniques used in assembling a producible definition of the product. This is the crucial element of the book. The information generated by these techniques goes beyond the current norm in product design. This information provides recognition of the physical form of the product and accounts for variation inherent in actual components. The techniques lead designers through the various phases of product development, refining the designs with geometric controls and enhancing producibility.

Chapters 7 and 8 introduce ideas and concerns associated with product verification. The order of presentation highlights the organization of the design cycle into phases that initially design the product, create the inspection system, and finally design the manufacturing process. The book's philosophy is that this order forces recognition of manufacturing reality. Once created in the first phase of development, the design should not be released into production if adequate verification processes cannot be defined. If it cannot be verified, the product definition is inadequate, inflating costs and possibly leading to an inability to manufacture.

Chapters 9 through 12 extend these ideas into the functional design of gaging, fixturing, and inspection processes. Sufficient examples are provided to demonstrate detail-level design techniques and to show how they are driven by earlier design phases. Engineers and designers will learn how their product design constrains tooling and gaging designs. Once recognized, these constraints can be turned to our advantage, allowing the physical product to be realized in an economically efficient manner. If done properly, development costs and engineering changes are substantially reduced as compared with traditional design processes in which variation is not directly anticipated.

The last two chapters, 13 and 14, consolidate the concepts into an actual case study and discuss implementation. The feasibility and power of the ideas are demonstrated by the case study, which serves to validate their use. In addition, with widespread interest in ISO 9001 compliance and certification in GD&T (ASME Y14.5.2), these chapters illustrate how structured product development based on GD&T creates an auditable process that adds value to the development cycle. Furthermore, it ensures that the benefits of this process are captured and documented, becoming a major asset and competitive weapon.

Note to the reader: figures in the book used to illustrate engineering drawings (hard copy or CAD) are not intended to be complete. On typical engineering drawings, the data density can be so intense that it is difficult to illustrate key ideas without forcing the reader to spend an inordinate amount of time decoding the drawing. Only after this task is accomplished does the reader arrive at the point where the real concept is apparent. To avoid discouraging people, the drawings have been simplified or abbreviated to a point sufficient to illustrate the idea but not to pass muster with the checker—if such a person still exists.

Also, it should be acknowledged that the term "drawing" is used to cover a variety of forms of engineering communication. We live in a world of mixed media where a drawing can be either a hard copy or in some electronic format; it is also increasingly likely that the product description may exist only in the form of a three-dimensional model with annotation attached in one of a multitude of ways.

I would like to take this opportunity to thank some of the individuals who made this book possible. In particular, I gained important technical insights from Al Neumann of Technical Consultants, Inc., who also serves on the ASME Y14.5M subcommittee. In addition, Sylvia Gurney deserves thanks for her efforts over the years that it has taken to bring the book to completion and for her aid in getting the manuscript ready to publish. I would be remiss if I did not thank Barbara Mathieu and Rita Lazazzaro of Marcel Dekker, Inc., for their guidance and patience in bringing this book to fruition. In addition, with the difficulty that normally attends reading technical books, the copy editor, Kristen Cassereau, should be recognized for the clarity she has added to the book. Finally, and most importantly, I thank my coauthor, Ed Roth, who provided the core of this book through his earlier works and who has continued his efforts to expose the benefits of GD&T to the design and manufacturing world.

Robert G. Campbell

Acknowledgments

We acknowledge the following workshops delivered by Ed Roth. The sponsoring organizations should be recognized for their support of the material and concepts developed in the book. Workshop participants contributed to the same effort by their questions and examples.

SOCIETY OF MANUFACTURING ENGINEERS

Design for Producibility, Dearborn, MI, August 1–3, 1994.

Functional Gaging and Inspection: The Database for Statistical Process Control, Schaumburg, IL, December 1–3, 1992.

Simultaneous Engineering: Linking Design, Process and Quality, Dearborn, MI, September 15–16, 1992.

Measurement Error Analysis: Techniques for Generating SPC, Albuquerque, NM, November 6–8, 1985.

Designing for Low Cost Manufacturing, Denver, CO, March 30–April 1, 1982.

Process Engineering for Manufacturing Managers, Dearborn, MI, June 12–16, 1967. Under auspices of ASTME.

UNIVERSITY OF WISCONSIN, MILWAUKEE

Geometric Dimensioning and Tolerancing, Milwaukee, WI, May 3–5, 1989 and May 8–10, 1989.

Product Design for Manufacturability, Milwaukee, WI, April 17–19, 1985.

NATIONAL SOCIETY OF PROFESSIONAL ENGINEERS

New Product Design, Atlanta, GA, October 1–2, 1981.

CREDITS

We gratefully acknowledge permission to reprint or adapt material from the sources cited below.

The following material is reprinted with permission of the Society of Manufacturing Engineers, Dearborn, MI.

Roth, E., Functional Gaging of Positionally Toleranced Parts, 1964, by ASTME as part of the Manufacturing Data Series, Library of Congress No. 64-23147.

Roth, E., Functional Inspection Techniques, 1967, by ASTME as part of the Manufacturing Data Series, Library of Congress No. 67-20359.

Roth, E., Gage designers can improve U.S. productivity, in *Gaging: Practical Design and Applications*, 2nd ed., E. Roth, ed., pp. 106–113, SME, 1983.

Roth, E. and Campbell, R., Geometric tolerancing, structure and language for concurrent engineering, *TMEH*, **6**, pp. 10-41–10-61.

Roth, E. and Campbell, R., Use of the ANSI Y14.5M standard in product definition, *TMEH*, **7**, pp. 14-1–14-18.

Contents

1

Introduction

In an interview given to *The Chicago Tribune* (Siskel, 1988), filmmaker Howard Hawks noted, "The one thing I've learned about making movies is that you can't fix a film once the shooting begins. If it's not right in the script, the problems are only bigger as the images move from paper to the big screen." The design and delivery of manufactured products are much like the making of a movie; both require equal sophistication and art for their execution.

Taking a page from Hawks' script, this book builds on the idea that once a product is designed it is too late to do anything more than put Band-Aids on production problems. A better way to get a product to market is to design it within a tightly integrated system and avoid these problems completely. Success comes from anticipating the interaction of the product design and the elements of production during the design phase when it is still economically feasible to make changes.

Previously such a design goal would have required the product engineer to be versed in design, engineering analysis, production techniques, quality assurance, reliability, and too many other disciplines to mention. The poor soul chosen to train in these areas would wind up with lots of gray hair and a short professional life. Furthermore, it would be dangerous for the company to concentrate this knowledge in a single individual when risks associated with the person's longevity, professional or otherwise, are considered. Of greater concern is the fact that allowing one individual to amass this proprietary knowledge makes it likely that it will not be formally documented and thus eventually lost. An integrated product development system provides the sole opportunity to eliminate or reduce such risks.

With the advent of low-cost computing power and relatively inexpensive networking software, integrating the phases of the development cycle can now be easily achieved. The design of a product can be approached systematically rather than through the piecemeal methods of the past. The challenge now is to understand the limitations of these systems. Product development organizations must realize that control is not always added by new management techniques and technology; rather, risks may be increased by using techniques that give a deterministic world view where one does not exist. Because the product's architecture is designed through a variety of technologically supported methods, different tools must be used in making the detailed decisions necessary to bring the product to physical reality. The responsibility passes from the "thinkers" to the "doers"—from the managers to the designers and engineers.

With this in mind, our book introduces a set of geometric techniques that can be incorporated within an integrated product development system. In fact, they are paramount to its success. The reader is given sufficient exposure to these methods to judge their validity and intelligently apply them. Each manufacturing situation is unique and requires creative and concurrent design of the product and the production system. As a consequence, the text is directed toward understanding the design phases of integrated product development. The resulting geometric viewpoint will allow the student or practitioner to create a product definition—encompassing both the product and the manufacturing processes—appropriate to the task.

An example helps to understand the magnitude of the problems that need to be resolved. One of the authors was hired to evaluate the design of a plain-paper copier when it was nearly ready for production release. A group of engineers from quality and manufacturing was looking at individual detail drawings for the first time, although the project was 30 months old. The engineers were dutifully signing off without the benefit of either a design layout or an assembly drawing. Without one or the other, no one had any way to determine the spatial relationships of the components.

These missing relationships are best illustrated by Figure 1-1, an exploded view of a similar situation that demonstrates the geometric relationships in a simplified assembly. Each of the components in the illustration has its own coordinate system (i.e., a datum reference frame). For both function and assembly, these coordinate systems must eventually be related or linked together and placed within a global system that defines the final assembly of the product. In the copier situation, this was not done.

Assuming for the sake of argument that all the rolls have sufficiently perfect

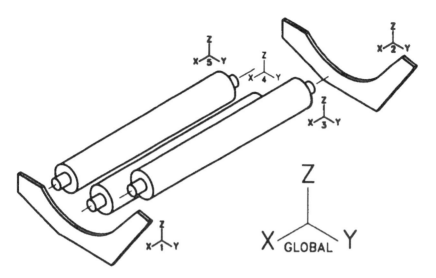

FIGURE 1-1 Illustrating a series of reference frames that must be linked for function and assembly.

geometric form to function properly, the spatial location of the roll axes can still defeat the design by being either mislocated or misaligned. The required relationships depend on accurately locating the mounting features in the frames. These frames are normally a series of mounting surfaces made from sheetmetal and plastic that connect to each other in a chain of subassemblies. In turn, the series of subassemblies mate to create the complete assembly. Any lack of geometric definition (e.g., axis location) can lead to large tolerance stackups, causing successful assembly to occur only by chance rather than by plan.

An examination of the engineering drawings revealed that the designs did not include any primary datums (reference features that aid in describing geometric location and relationship), which were necessary for an adequate review of the detail designs and an understanding of the desired geometry. This was astounding given the fact that the engineers had used the ASME graphics standards to define the product. More than 100 instances were related to applications of geometric control where primary datums were required but not provided. The specification of a primary datum was absolutely necessary in each of these instances to ensure functional conformance.

The preliminary analysis also indicated that the parts were dimensioned in such a way that their location was not defined relative to the copier drum. The engineering release was stopped and seven drafters were put to work locating these parts. From the information originally provided, five of the parts could not be positioned in the assembly because of implied, incomplete, or conflicting datums.

Several months later, after all the existing managers responsible for the components had been replaced, the project was out of control and the firm contacted a Japanese company to redesign and manufacture the copier. The marketing window was missed by two years, but the company finally got to put its name on a copier. This is only one instance of office products this division designed over a 10-year period—and all of them were found to be unmanufacturable. With a success rate like this, it is easy to understand why the division no longer exists and why thousands of employees in a small town lost their jobs.

To avoid similar problems, this book presents a number of tools that have been around for many years, at least in simplified form. Nothing illustrated here is based on radically new theories or technologies. Instead, the book offers a series of simple, yet elegant, methodologies that complement the best design and organizational techniques. Without straining credibility, these design methods will support any of the existing or likely techniques that management gurus offer. Be it total quality management, team building, design for assembly, agile manufacturing, or any of the myriad of ideas that come and go, the fundamentals required to deliver a product to the customer do not change. At each stage of the design and manufacturing processes, these basic steps must be undertaken. This book provides a mechanism to efficiently get through the cycle for a specific project.

The book is founded upon the concepts in the ASME Y14.5M dimensioning and tolerancing standard but goes beyond the usual drawing annotation applications. It specifically shows how these techniques may be extended to create elements of a design methodology. This methodology can then be employed in the simultaneous design and definition of a manufactured product and the processes necessary to produce it. The benefit of the standard is that it forces recognition of the variability of the design and production processes much earlier in the cycle than is currently done. The result is that function, durability, manufacturability, assembleability, and cost are considered throughout the development process.

Additionally, the book advocates that the standard be used as the primary medium of communication. Used in this manner, the standard links the various members of the concurrent engineering team, allowing them to communicate the unique definition necessary to transform the product concept into an economically effective production system. All subsequent product and process decisions are derived from this definition. The communication techniques and the organizational structure are intimately intertwined and are the key to project success.

The effective application of the methodologies offered here requires knowledge of the Y14.5M standard. This book is written in as general a fashion as possible to allow the presentation of these concepts to a larger audience, but extensive study and experience are needed to reap the vast economic benefits that can be obtained by their use. The book does not provide a detailed exposition of geometric controls. It does show, particularly through example, how the integrated approach coupled with geometric control may be used to add strategic value to product development functions.

In writing the book, the authors have drawn from the sections of the standard that support state-of-the-art manufacturing with an emphasis on single-setup processing. The resulting techniques provide a complete and uniform definition that is the hallmark of a well-engineered product development system.

Since existing design systems rely on the specialized knowledge of individual practitioners, it is necessary to look for a mechanism—a structure—that will serve to integrate the various disciplines (i.e., cross-functional teams) and still accomplish the goal of designing the product and the process simultaneously. The advantage of incorporating this structure in a team-based environment is that it leaves the individuals comprising the system free to retain their professional identities. This eliminates the inevitable compromises that will occur in the development of cross-functional individuals capable of dealing with these disparate disciplines.

Subsequent chapters describe the framework of the system and develop, through example, the various techniques that are its logical extension. Where possible, actual examples are used to illustrate basic concepts and show that they can be successfully implemented when applied in the industrial environment. What follows will provide a valuable strategic tool for all product development organizations. Be forewarned that the material requires conscientious study and continuous application for success. The steps in this book are but the beginning of a learning process that never ends.

REFERENCE

Siskel, G. Screen Gems Film's Finest Reveal the Secrets of Their Success, Chicago Tribune, 4 December 1988, Final Edition, sec. C, p. 32.

.

2

What Are the Techniques?

2.1 INTRODUCTION

The world of manufacturing thrives on continuous engineering change; only those firms that can react to these product changes and quickly fulfill the demands of the marketplace survive. To meet such a challenge, a variety of tools are necessary. This chapter highlights two tools being advocated and shows how they can be woven into a design methodology that reduces the total cost of product development.

The first of these tools involves the early and complete definition of the product and its supporting manufacturing processes. This requires a logical and disciplined approach to the design of the product and its allied systems if product development is to approach more of the science than the art of manufacturing.

The second tool concerns the organizational structure, a team-based environment, required to both provide and implement the product definition. Without this organizational foundation, only rare firms would have individual employees with sufficiently broad ranges of knowledge capable of using the methodology. Absent such individuals, the design process either would take too long to be successful in the marketplace or would not yield an adequate product design if done in a timely fashion.

These tools address a pair of distinct concerns related to product design—one of a technical nature and the other organizational. Both tools must be implemented simultaneously to achieve the desired ends. In all likelihood, the separate implementation of either one will not achieve the anticipated effects and may possibly make the product development process even more inefficient.

2.2 PRODUCT DEFINITION

Early in the historical development of interchangeable manufacturing, the end product was not completely defined in any formal manner prior to deciding what manufacturing methods would be used. Reducing engineered items to a precise graphical form (including tolerances) as a separate phase preceding production is a relatively recent occurrence (Ferguson, 1992) that can be traced back only to the latter part of the 19th century.

This was particularly true under the "contractor" system of manufacture (American Machinist, 1978) where the contractor was responsible for the design, manufacture, and delivery of the product. With this method of procurement, it was not unusual to find both design and manufacturing skills embodied in the same individual, as is still the case in Japan and Europe. The contractor would have anticipated many of the problems encountered in the manufacture of the product. As the contractor gave the product its final definition, he or she was also anticipating the design of the manufacturing system. Problems that would normally arise when responsibilities for design and manufacture are lodged in different individuals would be reduced or eliminated. The contractor, who had an immediate economic stake in the overall system design, was one of the earliest practitioners to apply concurrent engineering to volume production.

The advent of mass production and formalized programs of engineering study led to the separation of the design and manufacturing functions. With the enhanced scientific base that was incorporated into engineering programs around the mid-20th century, the severing of these functions was almost complete. Many of the current organizational management techniques, in both the human resources area and financial management, create the apparent need for this type of specialization.

It has been the authors' experience that disciplinary-driven walls exist in most firms, separating the design and manufacturing departments. Where this is the case, parochial and political interests prevent either of these disciplines from intruding into the other's area. The present organizational structure of many firms increases this isolation and makes it difficult to optimize the design of the product, or its production system, early in the design cycle. An additional factor that serves to reinforce this tendency is the serial nature of the design process as it has traditionally been practiced in the United States. The work of each functional department must be completed before the inputs to the next functional area are prepared and released. As a consequence, almost all stages of product development are critical in nature.

With the major design tasks on the critical path, the underlying design process becomes iterative and its elements interdependent. There is a constant give-and-take in the design of the product that affects previous decisions and decisions yet to be made. Once a preliminary design has been created, seemingly minor engineering changes can have a major impact on other elements of the product delivery system. This presents a multitude of problems, many resulting from the nature of existing organizational structures and reward systems that do not encourage dynamic interaction between design and manufacturing.

Thus, education, training, and organizational techniques have isolated many departments, making it difficult for them to communicate with each other. This lack of clear communication has particularly dire consequences when it involves the design and manufacturing departments. The further one is in the development process when design and manufacturing begin true exchanges of information, the more expensive implementation of the design becomes. This increase in expense can be expressed monetarily and in terms of time-to-market.

In creating any product, the principal criteria used in decision making involve the functional requirements necessary to satisfy market needs. These requirements can range from the utilitarian to the cosmetic. The specific techniques used to create the final product are all derived from functional specifications. One goal of the engineering process is to communicate the design intent embodied in these functional specifications. This communication should contain sufficient detail to ensure that any decisions made in implementing production will maintain the integrity of the envisioned design.

Current practice relating to the design of interchangeable products begins with the creation of concept layouts to provide designers, drafters, and management with graphical methods of visualizing the proposed product. As the design cycle proceeds and the preferred design concepts are chosen, a preliminary design layout is created that begins giving definition to the functional features of the assembly. As production is approached, the design layout is transformed into a detailed representation of each component that comprises the product. Throughout this process, the original design intent must not be obscured by the communication and documentation techniques or by alternative design interpretations introduced by these techniques.

Another factor that compounds the difficulty encountered in product development is the specialized knowledge lodged in many of the firm's functional departments. The parochial interests of these functional areas may serve to discourage rather than enhance both communication and recognition of the common organizational goals. The typical consequence is to take manufacturing out of the loop at the most critical phases of product design.

Until World War I it was common practice to use models of a product or its components produced by individuals with both design and manufacturing skills (Hounshell, 1984). A model could serve both as a prototype that would lend itself to experimental purposes—functional and process related—and as a definition of the desired end product. However, while it was advantageous to have this physical model available, the technique engendered some disadvantages as production became more decentralized with design and manufacturing separated by both time and distance. The model would now have to be replicated with a high level of precision and accuracy for use at a remote site. Any alteration to the model (i.e., the product definition) involved expensive and time-consuming changes.

When contrasting more recent use of working drawings to specify the product with an earlier reliance on a physical model, it is apparent that the use of graphical techniques to communicate design intent has a major impact on delivery of an acceptable product. The very nature of the physical model required consideration of manufacturing methods when fabricating this communication artifact. Functional concerns were automatically taken into account while creating the model. This method of documenting a component's design dramatizes the difference between the perfect part, created in the designer's mind, and the actual parts resulting from production.

The contrast between the graphical model and the fabricated part has a parallel in engineering analysis. The initial step in performing a mechanical analysis on an engineered part is to determine the force loading that may be applied to the part. This first level of analysis involves the use of statics, where the component being analyzed is assumed to be rigid; no deflections are assumed to occur on the part under the loads it is to withstand. Once the analysis is performed, the next step is to refine the assumptions made for the static analysis. This involves an elastic analysis that assumes small deflections occur. Resulting deflections and stress levels may then be calculated from the loads developed in the first-stage modeling.

The product design developed during the creative stage of the process is analogous to the static model defined in engineering mechanics. It is graphically modeled with the assumption that all the necessary features are perfect in form, orientation, and location. Such assumptions are inherent in the initial stages of the design process, whether performed manually or by a computer. Refinement of this ideal model occurs through an analysis that assumes deviations from perfect form, orientation, and location. This is analogous to the elastic analysis used to obtain deflections and stress levels. But the important issue is the measurable difference between the graphic model of the component and each physical part produced. Depending on how robust the design is, this deviation could inhibit or alter the desired function of the product. One of the designer's goals is to anticipate these interactions during the initial stages of design.

Graphical techniques that do not define products with datums and geometric controls cannot document and communicate design intent. Conventional design techniques, many of which are currently perpetuated by computer software, certainly do not stress any linkage between the product and the manufacturing system. If these design techniques do not incorporate geometric controls, then the physical variability resulting during production is not recognized and addressed. Only through the use of structured techniques based on internationally recognized standards can this information be attached to the graphical model in a way that will communicate the functional relationships that must be maintained during manufacturing.

2.3 THE LANGUAGE OF CONCURRENT ENGINEERING—Y14.5M

With market emphasis on the concepts of quality and value, a common language is needed to communicate the functional requirements of the engineering design. Of greater benefit would be a language that assists in establishing these requirements. The various individuals and departments involved in the product development process could all be versed in the use and interpretation of this language. As a result, they would arrive at a single specification of the product and the processes that generate it. The most important purpose of using this language would be its capability of giving a complete and unique definition to the product.

The best available tool to communicate the product definition is the ASME Y14.5M standard (and equivalent ISO technical reports and standards) that deals with dimensioning, tolerancing, and indirectly, metrology. For manufactured products requiring assembly, no other technique aggressively organizes the thought processes required in the design of the manufacturing system or supports discussion of the salient points as the product development cycle proceeds.

While this standard specifically applies to engineering drawings of products, fixtures, and gages, much of the material covered in the standard has even greater implications for integrating the various aspects of product design. The beginning pages of the standard contain a list of fundamental rules (ASME Y14.5M-1994), a number of which form the basis for this engineering language and are shown in Table 2-1. While this is an abridgement of the list contained in the standard, these seven items are key elements necessary to adapt the standard to the needs of product development.

An additional point to be made is that the application of these rules also forms a general design strategy that, in the strongest form, could be said to dictate the design of both product and process. To accomplish this, item 4 in Table 2-1 must be expanded to include the "datum reference frame" concept discussed in Chapter 3. With this element included, the concepts derived from the fundamental rules can provide structure to the geometric design process. No other techniques are currently available to integrate the often opposing needs of product designers and production personnel.

TABLE **2-1**   Fundamental Rules for Dimensioning and Tolerancing in Product
Development

1. Each dimension shall have a tolerance.
2. Dimensioning and tolerancing shall be complete so there is full understanding of
   the characteristic of each feature.
3. Each necessary dimension of the end product shall be shown.
4. Dimensions shall be selected and arranged to suit the function and mating relationship of a part and shall not be subject to more than one interpretation.
5. The drawing should define a part without specifying manufacturing methods.
6. Unless otherwise specified, all dimensions are applicable at 20°C (68°F).
7. All dimensions and tolerances apply in a free state-condition.

Abridgement from ASME Y14.5M-1994.

While engineers and designers commonly wish to conform to these fundamental rules, it is apparent that this happens to a lesser degree than is desirable.
The creation of a design strategy incorporating geometric control takes the first
critical step in providing a single design definition capable of reducing product
and process changes late in the manufacturing cycle.

Looking at the items listed in Table 2-1, particularly note the implied interactions. In existing applications of tolerance analysis it is assumed that the features to be analyzed are of perfect form and that the tolerance analysis proceeds
to deal with location variation on the basis of conventional techniques. These
techniques may deal with nothing more than two-dimensional, rectangular tolerance zones, possibly ignoring composite effects or problems of interpretation that
can lead to ill-defined zones and resulting assembly difficulties.

The rules recognize that all features are subject to variation not only of
size but also of form, orientation, and location. This set of defined variations
(positional variation is a prime example) can take forms other than those implied
by conventional techniques. A complete product definition not open to interpretation requires recognizing the interaction between these specific (i.e., three-dimensional) sources of product and process variation. More important, the standard
emphasizes the functional and assembly relationships that can only be controlled
through the refinements described by geometric controls.

By introducing this regimen into the design cycle, the designers can assess
and set product tolerances that are mathematically justified, not educated guesses.
Without considering geometric variations that will be experienced in production,
the normal approach to setting process characteristics is to fearfully err on the
side of tighter tolerances, due in part to the lack of manufacturing input into the
design. Implicit in this less-than-efficient technique is the knowledge that these
controls can be relaxed in the event that manufacturing encounters problems in

maintaining the specified levels. The product definition is no longer the inviolable source of design and process integrity.

Communicating by means of the standard returns integrity to the product definition and the development process. Many of the concepts embodied in the standard are refinements of ideas that have long played a role in the fabrication of interchangeable product (Buckingham, 1941). The ubiquitous datum reference frame and the concern with variation in component profiles are just two of the examples that can be used to illustrate historical continuity. The standard enabled product definition techniques to be codified in a way that now makes them understandable by both engineering and production disciplines. Communication is fostered at appropriate points in the design and manufacturing cycle where discussions and resulting decisions can impose discipline on a very creative, and often chaotic, process.

2.4 CONCURRENT ENGINEERING

The concepts found in the Y14.5M standard are the glue holding the concurrent engineering team together. Within the context of this book, concurrent engineering is the simultaneous design of the product and its allied manufacturing system. The scope of the system extends from marketing to the delivery, servicing, and disposal of the product, a scope sometimes referred to as "concept-to-customer."

With so many disciplines needed to support such a broad system, it is impossible to expect a single individual to anticipate all the important strategic and tactical decisions required to create a product. It is only through a team-based management approach that the design can be guided to a flawless and timely delivery. Concurrent engineering requires altering the more traditional, sequential project management techniques; responsibility and authority must be delegated and functional personnel properly allocated to the concurrent engineering team.

Figure 2-1 shows how substantial investments in intellectual assets begin to accumulate as the product cycle proceeds from concept to implementation. In lock step with the intellectual assets, physical assets such as hard tooling also begin to accumulate. As the cycle moves through the various phases that will be described in this book, a product definition is acquired in incremental fashion. The graph illustrates a varying lag between creating the product definition—at this stage more documented specifications than the complete understanding needed to produce the product—and actual knowledge of the product. The product definition only becomes manifest in the system as attempts are made to implement the definition and new knowledge is gained. Convergence of the definition and knowledge occurs only in the downstream phases. It is in these later phases that investments cause the cost of engineering change to rise in the nonlinear

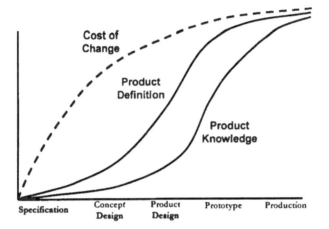

FIGURE 2-1 Effect of life-cycle phase on cost of change, product definition, and product knowledge. (Courtesy R. D'Alesandro, EDS PLM Solutions.)

fashion shown in the graph. Changes in choice of process, manufacturing equipment, or hard tooling can all incur tremendous financial consequences. In many instances, the expected costs of such changes will make any improvement in the product or process design impossible. Furthermore, while the financial losses attributable to these engineering changes may inflict considerable damage to the project, a more insidious loss is created by the increase in time necessary to bring the product to market.

The goal of the team is to provide a product definition that passes through the various phases of design, achieves design release, and then requires no further engineering changes. The team (Figure 2-2) should include all the members of the organization that have the required expertise, based on previous product development experience, to anticipate areas of concern and the necessary authority to resolve the resulting issues before design release.

While not intended as an exhaustive list of the members of the concurrent engineering team or as a treatise on team building and organizational dynamics, the following areas are identified to indicate the breadth of organizational talent needed to achieve definition of the geometric design. Additionally, suggestions are given as to how the Y14.5M methodology pervades the team's activities, enhances communication of critical concepts among its members, and increases the efficiency of the development process.

Team Leader. In a perfect world, the goals of the individual and the firm would coincide. However, it's not a perfect world, and there are usually parochial interests and turf battles with which to contend. Without a strong and impartial

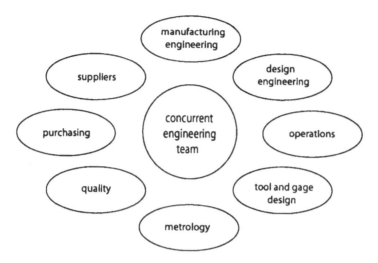

FIGURE 2-2 Composition of concurrent engineering team.

team leader and the appropriate organizational structure, many benefits of the concurrent engineering approach can be wasted as team and individual goals diverge.

Clark and Fujimoto (1991) contains a particularly good discussion of a variety of project management techniques that have been instituted in the manufacture of automobiles. Of particular interest is the experience related to firms that utilize a strong project manager who is given both the responsibility to set and the authority to achieve the project's goals. The use of matrix management techniques that do not give the leader sufficient authority, or that do not tie the individual's success to that of the team, is less likely to produce products that are competitive in the global market on the basis of either quality or timeliness.

Succinctly, the team must be composed of individuals whose personal and professional success is tied to that of the project. The team members must not act as parochial representatives of the functional departments. The preferred line of reporting for the team members should be within the organizational structure of the project and to the team leader. Dotted-line responsibility to the project and direct lines back to the functional departments are likely to ensure failure of the project team to meet the expected goals.

Marketing. In all product development activities, the direct needs of the customer must be considered. The inclusion of the marketing arm of the firm provides this representation, ensuring that, as the product development cycle proceeds, any technical decisions that are made keep the consumer—the ultimate judge of the design's success—in view. The marketing representative assumes the role of the consumer's advocate.

In defining the architecture of a technically complicated system, tools such as quality function deployment may be used to interject the customer's voice at the front end of the project. The marketing representative can provide this impetus in a manner that enhances the likely success of the product.

Design Engineering. As expected, design engineering and its co-leader, manufacturing, assume a major portion of team responsibilities. Design engineering is arguably the discipline most in need of the leavening effects brought to the table by the concerns of manufacturing. It is also the discipline that may gain the most from the structure imposed by the Y14.5M regimen. Since the latitude available to the downstream elements of the production system is circumscribed by the product definition, it is useful to again point out that the standard is intended to cover the exercise of dimensioning and tolerancing methods as they relate to the product, fixture, and gage definition. Both design and manufacturing engineers must take the lead in expanding the use of these tools to achieve a viable production system. While the designers are ultimately responsible for justifying the type and level of control for the product, they must realize that the results of such an analysis will not be neat numbers (e.g., the same tolerance values for all analyses). Also, there may be considerable negotiation with manufacturing and other functional disciplines before successful implementation. The design engineer's choice of controls influences and possibly constrains the design of process elements.

Placing design engineering near the top of the list is intended to illustrate the importance of the design effort to subsequent attempts to control costs and implement process improvements. From studies conducted at Ford, approximately 70% of the product costs are locked in during the design stages (Boothroyd et al., 1991; Nevins and Whitney, 1989). In the traditional design regimen, completion of the design severs the designer's major responsibility for successful product delivery and, as a result, solidifies the cost structure. Any design engineering involvement past this point is based on exception. Larger organizations will have formalized engineering change request systems. These requests usually flow to the engineering department, where they will be evaluated and changes effected only if appropriate criteria are met. However, design engineers typically do not see direct involvement in process development as being within their job description. Of greater importance would be the resistance to suggested changes by design personnel exhibiting the "not-invented-here" syndrome. This last item is a very real possibility where design engineers are asked to evaluate either product or process improvements initiated by other disciplines.

Because of the design's fundamental impact on product cost, the group responsible for the product's physical definition should lead subsequent development and improvement activities. There is a major advantage in placing design in this key position. Design engineering provides an institutional memory by documenting the results of the design process and incorporating them into the information stream. This prevents design information—intellectual property—from slipping away and ensures that the results are used in subsequent implementations.

In any design there are tradeoffs associated with the choice of geometric controls and component producibility. The product designer's choice of an appropriate level of control should be the result of a formal analysis, and not a default to standardized levels. For example, Figure 2-3 shows the range of controls that might result from analysis of a mechanical fastener mating with a clearance hole. The criterion used to assess the alternatives was that the fastener's head covers a clearance hole or slot. Each of these choices has a different impact on producibility and, ultimately, on cost.

Manufacturing Engineering. The physical realization of the product depends on the ingenuity of manufacturing engineers. Critical choices pertaining to manufacturing processes and functional concerns merge in this department, ultimately creating the product's cost structure. The resulting decisions move the

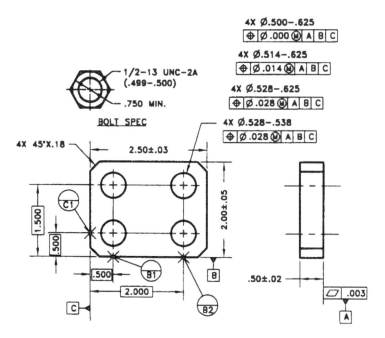

FIGURE 2-3 Alternative callouts for positional tolerance.

process beyond the original description implicit in a preliminary product definition and into the physical realm. Many of the opportunities that add further value to the production process surface through the comments and concerns of manufacturing engineering. In particular, product variation represented by tolerance specifications does not manifest itself until the product assumes physical form. Logically, the greatest concerns relating to variation are within the purview of the group giving the product reality.

Manufacturing process knowledge is incorporated into the design phase whenever geometric control is used. The underlying standard contains elements found in the language and knowledge base of manufacturing engineering. Such knowledge is of particular importance to individuals whose education and experience do not encompass sufficient manufacturing background for them to truly appreciate functional variation on actual components. Prior to actual production, the basic concepts in the standard may provide the only mechanism able to trigger an investigation by an inexperienced designer of these process variations.

Another point that should be considered is that the concepts found in the standard have been incorporated into manufacturing science during its entire history. Many of the concepts and controls that the standard identifies are commonplace in the manufacture of discrete products. The introduction of reference surfaces (a datum reference frame) to qualify a component prior to fabrication is just one case of manufacturing techniques being codified in the standard. In cases where the simultaneous design philosophy is not followed, the ultimate responsibility for the final design of the product rests with the manufacturing group; they are the ones who must get the product out the door. To avoid deferring final design decisions to manufacturing, Ford reorganized in 1985, placing both design and manufacturing under the same vice president, thus ensuring cooperative concurrent engineering. Another example involves the Department of Energy weapons design labs, where the ASME Y14.5M standard (and its predecessors) has been used for over 30 years in defining products, fixtures, and gages.

Operations. As the text alludes, many of the opportunities affording the greatest leverage in the concurrent engineering process are to be found in the way the firm is organized. Two specific concerns have to be addressed: the organizational structure and the organization of the work. The importance of placing an operations representative on the team is to affect the way work is accomplished. Massive inefficiencies can be introduced into the fabrication of a product by the inappropriate application of materials, processes, and work methods.

Work methods are singled out to emphasize the interaction between technological tools and production personnel. Successful improvement of processes used to manufacture a product occurs only through the effort of the usually unnamed production workers who deal with many of the day-to-day problems. These problems are the result of material variation, manufacturing environment, and a myriad of other influences that can never be entirely anticipated, even by the most successful product design organizations. Production personnel must actively participate in process design and improvement as members of the concurrent engineering team.

A further consideration is that any projects originating in other functional areas will likely have considerable impact on the operations group. Operations must be an integral part of the process that defines the work they will be expected to accomplish. Their full cooperation can be garnered by allowing them to be part of the design and improvement activities of the team.

Tool and Gage Design. With increasing emphasis on total quality management, the inclusion of the individuals who will be responsible for designing much of the process hardware is paramount. From the specification of process tooling and gaging to the design of custom tooling and CMM fixtures, the tool and gage designers heavily influence the level of manufacturing efficiency.

Much of the fundamental activity of tool designers starts with the positioning of the component in three-dimensional space. The method used to accomplish this is normally referred to as the *3-2-1 location technique*. While not described in tooling texts in the specific language of the Y14.5M standard, this is a detailed application of the datum reference frame concept covered in Chapter 3. The designers can reduce setup error and consumption of the working tolerances by identifying and qualifying datum features and specifying geometric controls when creating the tooling and gaging devices.

In the situation where dedicated tooling is to be included in the manufacturing system, many of the functional characteristics identified as crucial to the product definition will influence hardware design. These concerns are communicated to the tool and gage designers by the datum reference frames and feature controls found in the product definition.

Metrologists. With the ever-present trend toward more precision and higher resolution in measurement, a member of the team must be versed in dimensional metrology principles used to establish product conformance. The presence of such an individual ensures the integrity of the quality system database utilized throughout the development cycle.

Employing a metrologist is also important to deal with virtual models of the product created through the use of inspection and measurement procedures that may include soft gaging (i.e., computer-based, computational methods). Many measurement techniques incorporated into computer-assisted equipment (e.g., CMMs) do not conform to the definitions—through either software implementation or actual practice—contained in Y14.5M, hence the development of mathematically rigorous definitions for geometric controls (ASME Y14.5.1). The metrologist will be expected to create techniques to mitigate these effects.

Quality Assurance and Quality Control. A representative of the quality assurance group is necessary to ensure the delivery of the product at the appropriate level of conformance. Implicit in this is the idea that product manufacturing is embedded within a larger system based on the principles of total quality management.

The responsibilities of the quality representative may be divided into two separate tasks. The first is quality assurance, explicitly stating the quality system requirements. This is a management and planning function that designs the system and provides the procedures necessary to obtain the desired product. Quality assurance also provides technical assistance to the team such as determining those product characteristics that require statistical charting to establish process capability and maintain control. Many of these characteristics are either datums or features controlled by geometric specifications. Additionally, quality assurance provides the technical expertise to deal with the types of experimental design used to increase the robustness of the product definition (product and process) and to ensure its completeness.

The second task is that of quality control, the actual execution of the quality activities needed to obtain the desired quality level. Ultimately, quality assurance concerns must be reduced to executable procedures and hardware. The quality control function, responsible for implementation, is represented on the team since it is its personnel who confront the actual variation, manifested in both the product and the measurement system. The team members should also be the skeptics questioning whether the critical characteristics that have been identified produce sufficient functional or process effects to warrant the expense of explicit control. Stories abound purporting to show use of statistical process control techniques applied to nonfunctional features only as a consequence of contractual requirements, creating a complete waste of inspection dollars.

With the product definition serving as the focal point for product and process design, it should be obvious that adherence to the definition must be verified. Achievable verification methods must be specified during the product design stage. This means that either specific variables or attributes must be identified and included in the definition as controlled characteristics. At a more general level, no product feature should be included if a metric does not exist to specify it and if an inspection method is not available to verify conformance. With the team's consensus, the quality assurance representative suggests the metric and its verification method. This argues for the use of the Y14.5M standard both as the medium of communication and as the means of specifying assurance and control levels.

Purchasing. With firms making strategic decisions to subcontract many of their manufacturing needs, purchasing must be involved in team discussions starting with the preliminary design stages. Vendor lists maintained by the purchasing agents enrich the knowledge base relating to process choice and provide a wider array of alternatives for use in the design process.

With design activity focused on creating a precise definition of the product, purchasing must be conversant with the Y14.5M standard and its ability to communicate design intent. The purchasing representative must impress upon the firm's suppliers the need to correctly interpret and apply the controls used to ensure product conformance. The purpose of these activities is to obtain greater value resulting from clearer engineering requirements and the use of techniques such as bonus tolerances allowed by the Y14.5M standard. The activities also help to identify verification methods to ensure conformance to engineering requirements. Such activities are most effective prior to the supplier's starting production since this may influence the choice of successful bidders and eliminate subsequent production problems. Y14.5M controls should reduce, not increase, the cost that the supplier will incur in performance of the contract, a direct consequence of efficient and concise communication of the engineering requirements.

Since firms rarely have complete vertical integration, an even more important level of involvement for the purchasing agent occurs when the company not only purchases completed components but also obtains technological knowledge bundled with the physical purchase. Using automotive firms as an example, many of the subsystems acquired for assembly into the completed automobile are purchased with component engineering included with manufacturing. It becomes crucial that the purchasing representative is involved in the subsystem development and conversant with standardized product specification techniques. Such involvement allows seamless integration of external resources that can augment a company's technical and engineering capacity. The purchasing agent may also act as a liaison between the company and its vendors, as suggested in the next section.

Suppliers. Again considering the large percentage of value that may be added to a product by external sources, it is important that members of principal supplier organizations be included on the team. They bring specialized knowledge to the discussions that greatly impact the design of the product or its allied processing systems. Benefits can accrue from the specialized knowledge and experience that the subcontractor has in either particular product types or special manufacturing techniques. In many cases, these benefits arrive without incurring additional cost.

Team participation gives suppliers an incentive to actively study the use of geometric controls. They cannot participate in the simultaneous engineering process without people who can interpret and apply geometric controls. The refusal of any organization to undertake this training may provide the host organization with an early indication that an alternative source is needed.

Normal supplier activities are directed toward process-related tasks and require performance based on legally executed contracts. Since the engineering drawings and CAD models (the product definition) form part of the contractual agreement, the individual suppliers can benefit from participation on the concurrent engineering team as the product definition is created. Benefits result from the supplier's ability to influence the design early in the cycle, where altering the definition of either the product or the processing system has less more economic impact. If the supplier gains more flexibility, then the product becomes more producible. This contrasts with the practice where a contract is based solely on solicited quotations obtained using rigid (and possibly incomplete) product definitions determined without manufacturing or supplier input.

When purchasing major portions of a product assembly, subsequent contractual disputes over product definition can be reduced by including process-driven individuals (the vendor's personnel) early in the cycle. Potential problems are eliminated by the early resolution of misunderstandings or disagreements relating to the interpretation of the product definition. Ultimately, this serves to reduce the total delivered cost of the product and meet delivery schedules.

This situation should be used for competitive advantage. The obvious way is to integrate suppliers as long-term members of the product design team. Involving the vendor in the design process also introduces the possibility that useful alternate technologies of which the product design team was not aware may surface. In-house technical talent usually has such a broad range of responsibilities that it is unlikely that they will have the depth of knowledge in specialized areas to foster consideration of alternate technologies. Wise choices in vendors and longer-term contractual arrangements can create tremendous technological leverage far beyond the internal capabilities that can be economically justified. The advantage to the purchaser is the ability to obtain state-of-the-art technology without the need to develop and maintain these resources in-house.

2.5 SUMMARY

For successful product development to occur, two specific tools must overlay a firm's product development practices: team-based concurrent engineering and complete product definition based on the tenets of the dimensioning and tolerancing standard. The latter tool is used to yield an early and complete product and process definition, extending from the actual product description to the processes required to produce it. When used together, they provide a design methodology that gives a competitive advantage in the product design and development area. The term "product design" implicitly includes the design of the manufacturing systems—manufacturing and inspection—along with the traditional understanding of the product as the entity the customer receives.

The book recognizes the reality of manufacturing; the product definition in the mind of the designer is not the actual product that emerges from the manufacturing plant. The techniques illustrated show how a complete and unique product definition can be created, taking the variation of real-world production into account. This unique definition serves to reduce product development cycle time.

The thrust of the remainder of the book is to show how the technical and organizational structures must be integrated to achieve the benefits of concurrent engineering, although the text places the greatest emphasis on specific technical methods. A complete product definition (including geometric controls) enables the concurrent design of the production and verification systems. The result is that the majority of product and process variation issues have been raised and addressed in an organizationally effective manner. Since the greatest portion of the product's cost structure is locked in once the product drawings are approved and released, the integrated approach eliminates many of the changes that might have been instituted late in the development process. Hence, basing the design process on collaborative efforts using the concepts of geometric control provides tremendous value to a product development organization. The leverage in time and dollars obtained by implementing such practices far exceeds the benefits to be gained from management fads that companies look to as quick-and-easy solutions to their problems.

REFERENCES

American Machinist, Metalworking: Yesterday and Tomorrow, New York: American Machinist/McGraw-Hill, 1978.

Boothroyd, G., Dewhurst, P., and Knight, W., Product Design for Manufacture and Assembly, New York: Marcel Dekker, 1994.

Buckingham, E., Principles of Interchangeable Manufacture, 2nd ed., New York: Industrial Press, 1941.

Clark, K. B. and Fujimoto, T., Product Development Performance: Strategy, Organization, and Management in the World Auto Industry, Cambridge, MA: Harvard Business School Press, 1991.

Ferguson, E., Engineering and the Mind's Eye, Cambridge, MA: MIT Press, 1992.

Hounshell, D. A., From the American System to Mass Production, 1800–1932, Baltimore: Johns Hopkins University Press, 1984.

Nevins, J. L. and Whitney, D. E., Concurrent Design of Products & Processes: A Strategy for the Next Generation in Manufacturing, New York: McGraw-Hill, 1989.

3

The Basis of the System

3.1 INTRODUCTION

If perfect parts could be manufactured, most design and production problems would disappear. What the designer conceives, manufacturing creates. However, in a less-than-perfect world, variation occurs and must be confronted. This begins early in the product development cycle and extends to subsequent continuous improvement activities. The reality of manufacturing depends on the identification and removal of common sources of process variation. The quest for the perfect product requires both an effort and a discipline that must become part of the manufacturing culture.

Based on the authors' experience, the product delivery system contains some predominate causes of common variation. The designer of the product, the tooling, or the measurement system creates this variation and its effects; the users of ill-conceived designs can only suffer in silence. The structured design techniques developed in the book are intended to aggressively address this ever-present variation.

The critical design source of variation is best described as the inept product definition that occurs when products are documented by drafters who operate in a vacuum and do not use the appropriate design language—the ASME Y14.5M standard.

The major manufacturing source of variation is multiple, nonfunctional fixtures (both tooling and gaging) that hold and present the part being manufactured in a manner that makes the features on the component difficult to relate to each other. The resulting errors propagate throughout the geometry, making the component troublesome to assemble and limiting its ability to function as intended. The existence of several process-related datum reference frames (discussed in the next section) on a single part is one indication that such nonfunctional fixturing may have been designed into the production system.

In the quality system, a similar source of process variation is measurement uncertainty. Much of this is again the result of designs based on nonfunctional gaging and CMM fixturing techniques. From the authors' experience, this element, in conjunction with setup error caused by nonqualified part datum features, can consume from 50% to 300% of the part tolerance. A control chart using such an invalid database will represent only measurement error, not the variation of the characteristic being monitored.

To bring the values of these common sources of variation within acceptable bounds, such that they do not consume a preponderate portion of the part tolerance, the following concepts are used to assemble a framework, both technical and organizational, on which to construct the integrated product development system.

3.2 Y14.5M CONCEPTS

To gain the benefits of geometric control, a number of fundamental concepts are used to build a structured design methodology. The most important of these ideas involves the use of datum reference frames. Four additional concepts are added to form the foundation of geometric control: These include interrelated features, the boundary concept, Taylor's principle, and refinement of controls. Each is a critical element in the creation of a system of design methodologies and geometric control. When designers do not have an operative knowledge of these elements and there is no established concurrent engineering team, an incomplete product definition results, ceding control of the product and its allied process design to individuals located downstream in the development process. These downstream "designers" now have the freedom but not necessarily the knowledge to make optimum decisions about product function; certainly, they should not be the ones to provide the primary functional definition of the geometric design.

3.2.1 Datum Reference Frames

Of the concepts embodied in the Y14.5M standard, the pivotal one needed to structure engineering activities is the datum reference frame (DRF). As mentioned, the designer has an idealized view of the results of the manufacturing process. The conceptual design of a product does not normally look into the vagaries of the physical world and their effect on execution of the design. Under normal circumstances, this type of information is usually encountered during the development of the production prototypes or the production system itself.

To ensure that the design intent is maintained, specific types of information must be generated as the preliminary design layout evolves into detail drawings. The dominant piece of information relates to kinematic control of a component, which establishes the relationships within the product assembly. In its most general form, kinematic control is created by a DRF describing how the six degrees of kinematic freedom (discussed ahead) are to be constrained. The DRF describes the three-dimensional location of the component, providing for subsequent assembly and function.

While it sounds complex, the idea of datum reference frames has been around for a very long time. The old machinist's dictum to fixture a part once and do all the required machining in a single setup illustrates this fundamental concept. Each time a component is taken out of one machine setup and placed into another setup (usually requiring another fixture), variation—or error—is introduced into the process. Each additional method of locating the part establishes a new set of reference surfaces, adds to the geometric mix of the part features, and causes tolerance stackups (variation) that play havoc with the functionality of the assembled product.

The DRFs contained in the product definition establish a unique location in space for the part based on the function of the component in the assembly. For relatively simple parts such as cylinders or prismatic solids, the DRF can readily be explained in terms of the Cartesian coordinate system. As demonstrated in Figure 3-1, an object has six degrees of freedom in three-dimensional space.

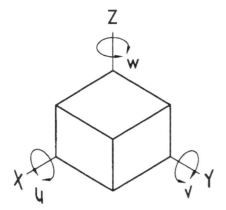

FIGURE 3-1   Six degrees of freedom for kinematic control.

Three of these are the motions it can have along any one of the coordinate axes. These linear motions are referred to as *translations*. In creating or assembling the component, these motions are usually constrained. Additionally, there are three further motions, *rotations* that can occur about the coordinate axes. Again, in the fabrication or assembly of the product, these motions are normally restrained although, for example, if the part is turned on a lathe or rotates in the assembly, not all the degrees of freedom are constrained.

The significance of the DRFs goes back to the machinist's rule-of-thumb. The minimum level of variation inherent in a design occurs when a single DRF is associated with an individual component on the design layout and this same DRF is used to manufacture the part. Additional DRFs used to locate the component during manufacture or measurement introduces greater feature variation, which can cause problems in achieving the desired part function. With increasing demands on part quality and significant improvement in machine precision, such variation is no longer acceptable. The use of the single setup (i.e., a single DRF) has again been discovered by machine tool marketers and has become a marketing aid to sell machining and turning centers. While a single DRF is not always obtainable, it does provide a benchmark toward which the design and improvement team can aim.

Examples of the two most general types of DRFs are shown in Figures 3-2 and 3-3. The first diagram shows the application of this concept to a prismatic part and indicates how the object is located in space using three orthogonal planes. The underlying theory (in heuristic form) requires that the object contact one of these planes (primary plane) with three points of engagement; two points of con-

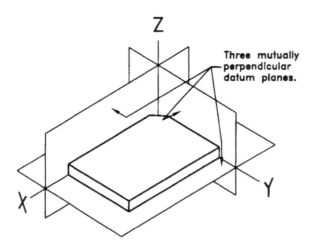

FIGURE 3-2 Prismatic part related to datum reference frame.

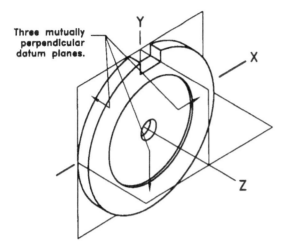

Three mutually perpendicular datum planes.

FIGURE 3-3 Cylindrical part related to datum reference frame.

tact establish the second plane (secondary); and a single point determines the third plane (tertiary). This technique is commonly referred to as the *3-2-1 method* and is discussed in more detail below.

Figure 3-3 shows how this concept can be applied to a cylindrical part. In this case the three-plane concept may still be applied. However, the location of the component is now determined by the coincidence of the line of intersection of two of these planes and the axis of the cylindrical feature on the part.

The processing equipment establishes the DRF for an actual part. In the case of Figure 3-2, the DRF could be the table of a machine tool that may simulate the primary datum, with rails providing the secondary and tertiary datums. The cylindrical part (Figure 3-3) would have its DRF determined by the chuck or mandrel on a lathe where the axis of the true cylinder (i.e., the ideal part) is simulated by the machine tool axis. The secondary datum (a fixed stop or friction) would prevent motion along its axis; the tertiary datum would constrain rotation about the axis and could be simulated by a stop.

The key pieces of information needed to define a particular DRF include the functional definition of the component as it relates to the next level of assembly and the spatial relationship between this part and the other components. Using these relationships, the design team establishes the theoretical datum planes appropriate to a specific design. By focusing on the functions and relationships in the assembly, the primary design intent is automatically considered during any discussions related to the component and its manufacture. The DRF allows us to visualize and describe the part's location in three-dimensional space, and it maintains functional integrity as the concurrent team continues its work.

When a simple prismatic part is designed, it is usually drawn in fixed relationship to mutually perpendicular reference planes; part features (holes, surfaces, etc.) are then located by means of dimensions from such planes. The reference planes may be compared with the *x*-, *y*-, and *z*-axes of the Cartesian coordinate system (Figure 3-4) and the resulting coordinate planes.

Part datum features suitable for contact by functional tooling, fixtures on CMMs, or gages (and, of course, by mating part features) are identified by the appropriate datum symbols. Part datum features should, of necessity, be more accurate than the features located from them; otherwise, setup inaccuracies (nonrepeatability) can rob the part features of a large percentage of their allotted tolerances. Accuracy ratios of 10 to 1 are common for traceability to the National Institute of Standards and Technology (NIST). Note that the 10 to 1 ratio might be relaxed in the event of tight tolerances. The high precision made necessary by tight tolerances could quickly exceed the measurement process capability if the 10 to 1 ratio was applied at each step of the path leading to the primary standard. This is a good argument to set these specifications based on analysis rather than relying on default ratios. To be useful for tooling and gaging purposes, dimensions from part datums should be directly usable without calculation. These are provided by "basic" dimensions, the perfect location based on design intent, which do not introduce tolerance stackups even when the dimensions are chained together.

"Datum" dimensioning (i.e., all dimensions are given within a single and complete datum reference frame) is entirely compatible with conventional machining practices and practical tool and gage design. As mentioned, it is highly

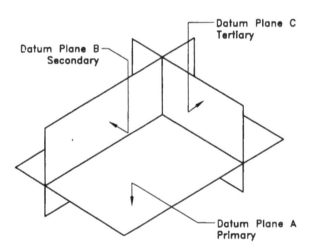

FIGURE 3-4 Datum reference frames.

advisable to set up the part only once in relation to the three mutually perpendicular machine tool axes and, if possible, to have subsequent tools and gages use the identical DRF if several operations or gaging steps are required. The use of chained (related) DRFs on an individual component does not facilitate the use of functional gages since multiple DRFs on the same part each require individual setups, individual tools, and individual gages. Unless the part is too large to process at one setup, chaining of part DRFs is not practical because of higher tooling and production costs.

Part Datum Features. Figure 3-5 shows a three-hole pattern dimensioned from three mutually perpendicular part datum surfaces, with each datum assigned a letter designation and its precedence shown in the feature control frame. Part datum feature *A* is the primary locator, the most important alignment surface, a plane established by its three highest points (Figure 3-6). Part datum feature *B* is of secondary importance, established by its two highest points. Part datum feature *C*, of tertiary importance, is established by its single highest point. The theoretical datum reference planes are shown in Figure 3-4 and their degree of precedence in part location and alignment is shown in Figure 3-6, which combines Figures 3-4 and 3-5. Location is determined by forcing the part into contact with datum reference plane *A* at the three highest points on part surface *A*. Then if the part is forced ("banked" in toolmakers' terms) against datum reference planes *B* and, subsequently *C*, surface *B* will contact datum reference plane *B* at

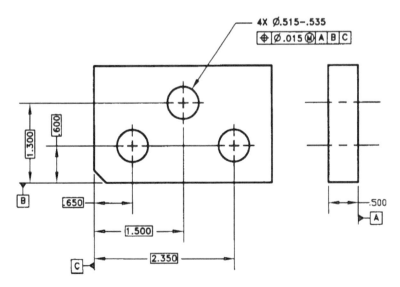

FIGURE 3-5 Example of geometric dimensioning using datums.

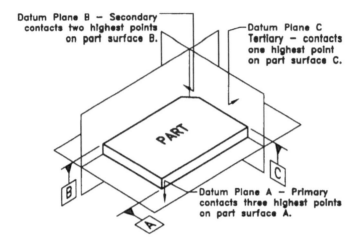

FIGURE 3-6   Part related to datum reference frame.

the two highest points; surface *C* will contact datum reference plane *C* at the highest point. The remaining degrees of freedom are removed by clamping devices, ideally placed opposite the locating points.

This desired locational relationship will exist when:

1. There are three mutually perpendicular planes (three machined tool surfaces will approximate this).
2. The part surfaces are not convex.
3. Solid three-, two-, and one-point contacts are ensured through the use of datum targets in the form of commercial locators (instead of machined surfaces on the gage) that contact the part at specified locations.

Figure 3-7 shows how the actual part shown in Figure 3-5 can be dimensioned from datum reference planes established in a different fashion than Figure 3-6. This is accomplished as follows:

1. The plane through the three high points on part datum feature *A* that has 0.001 flatness tolerance—this qualifies the datum feature and stipulates that it cannot be convex (which would allow the part to rock on a machined tool or gage surface).
2. The plane through the two points on part datum feature *B*, the secondary datum, with the tool or gage pickup points (datum targets) illustrated.
3. The plane through the one point established by the tool or gage pickup point (target) shown against part datum feature *C*.

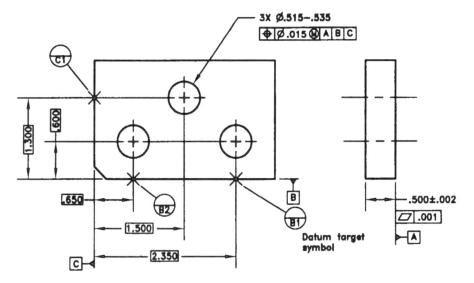

FIGURE 3-7   Geometric dimensioning using datum targets.

3.2.2   Tooling and Gage Datum Elements

Actual datum planes (called ''datum features'' to distinguish them from datums, the origin of all measurements) are determined by the geometry of the functional tooling or gage datum elements. The actual datum features described in theory by the datums shown in Figure 3-7 are established by the tool or gage shown in Figure 3-8. Part datum feature A is contacted by a machined surface (i.e., datum feature simulator) on the tool or gage, and part datum features B and C by three dowel pins (datum targets) that further locate and align the part.

Figure 3-9 shows part datum features symbolically contacted by six tool or gage pickup points, datum targets that are particularly useful on cast and forged parts. The datum targets are placed on part drawings with basic dimensions (or with conventional tolerancing) subject only to gagemaker tolerances. The resulting tool or gage configuration consists of three toolmaker's buttons (areas) that establish datum reference plane A, and two tangent (line) locator elements that establish datum reference plane B, and a tangent locator element for the datum reference plane C. In theory, these locators could all be point locators.

Figure 3-10(a) shows how a planar datum and datum targets combine to describe a different tool than that shown in Figure 3-9. The tool-locator symbols—which would be targets on the actual drawing— placed on (or straddling) the part contour in Figure 3-10(a) and located in the adjacent view correspond to datum target positions described with basic dimensions to indicate the position

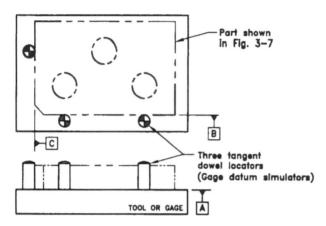

FIGURE 3-8 Fixture using combination of datum simulators.

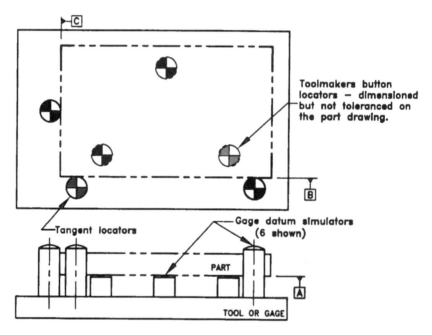

FIGURE 3-9 Fixture showing combined tangent and button locators.

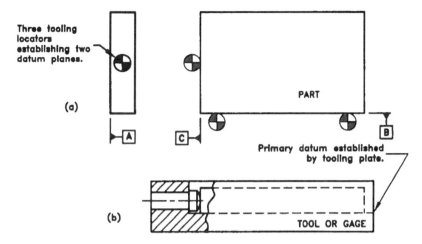

FIGURE 3-10   Fixture showing use of button locators and tooling plate.

of tooling elements. These are shown in the tool or gage configuration in Figure 3-10(b). Note that the use of the planar datum feature rather than datum targets now requires the use of the tooling plate from which the three points of contact must come. This is a larger set of points than those contained in the reduced point set described by the targets.

Figure 3-11(a) shows the alignment that could occur if part datum feature A contacts a tool made up of datum elements consisting of tangent and button locators. Figure 3-11(b) shows the identical part placed in a gage consisting of tooling plate and rail datum feature simulators; Figure 3-11(c) shows that the latter design will not align the part in the same manner as did the tool containing the six targets. These setup errors can occur with respect to all three datum planes and may reject parts because the same reference plane is not established in each case as the tool or gage design changes. Perfectly square and flat part datum features would not result in any misalignment if different tool and gage designs were used; but since the perfect part does not exist, it is more practical to specify the actual datum reference planes used by defining the tool or gage datum element geometry. When the part datum features firmly contact the tool or gage datum feature simulators, the actual datum feature for manufacturing and measurement is established. In surface-plate inspection, the geometry of angle plates, close-fitting pins, and the surface plate establish simulated datum elements that, when contacted by part datum features determine the actual datum features for measurements. The use of identical tool and gage elements ensures that the least amount of tolerance is "robbed" from all features dimensioned from the part datum surfaces. Both the functional tool and the gage should simulate the mating part where

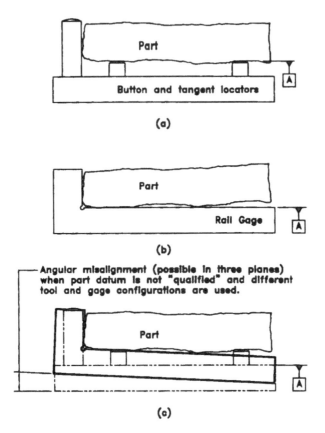

FIGURE 3-11   Possible setup errors.

they contact the part datum surface if at all possible. This simulation will reduce setup errors because (1) it is both practical and natural and (2) it will deform the datum feature on flexible parts no more than the mating part at assembly.

To summarize, part features are dimensioned on engineering drawings from *part datums* that are perfect in the designer's mind; they are manufactured creating *datum features* and measured from reference planes associated with *datum feature simulators* that are real surfaces on fixtures.

3.2.3   Interrelated and Interchangeable Features

Many of the products available today achieve their value by taking individual components and creating an assembly. With an assembled product comes the desire for component interchangeability. To achieve this, the product designer

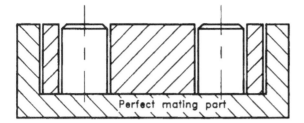

Perfect mating part

FIGURE 3-12   Conventional control showing perfect form, orientation, and location.

and manufacturing engineer must consider the relationship of the various features on the individual components and the relationships between mating parts.

No existing product definition technique can communicate interchangeability criteria, with the exception of those codified in Y14.5M. Using conventional dimensioning and tolerancing schemes based on this standard, a unique interpretation of the drawings is impossible. In the latter case, the drawing user would inevitably conclude that there are no explicitly stated interrelationships of size, form, orientation, or location necessary to allow assembly of the component. The designer's vision of the part function is not communicated. If such relationships are desired, then the Y14.5M techniques must be explicitly used. Nothing can be left to the imagination.

As an illustration of these relationships, Figure 3-12 shows a part that is to mate with a second component that contains a number of pins. As the surface of the mating component carrying the pins engages the part, the fixed pins must simultaneously enter the holes on the mating part. These pins enter the mate only if attention has been paid to the relationship of the holes to the surface they are mounted in, to each other, and to their location. As shown in this figure, perfect size, form, orientation, and location have been assumed, the first level of analysis that a designer might contemplate.

Another view of the world would specifically recognize variation resulting from the actual production process. Specific limitations on relationships that result from the actual process would be precisely communicated by having the designer specify a DRF and apply geometric controls. These techniques can now be used to determine the extent of allowable variations in the geometry and their effect on the assembly process (Figure 3-13).

3.2.4   Boundary and Axial Concepts

When analyzing an assembly for tolerance stackups, it is helpful to visualize the worst-case assembly condition (the virtual condition envelope). This envelope, or boundary, aids early consideration of the type and level of tolerance control

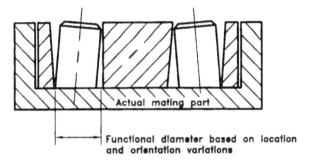

FIGURE 3-13 Geometric control showing possible variations in form, orientation, and location.

that must be maintained to produce a functional product. Using positional tolerancing as an illustration, the standard introduces two ways of looking at the tolerance zone.

The first method identifies a bounding surface (e.g., the virtual condition boundary in Figure 3-14) that defines the cumulative geometric effect of the controls applied to a specific feature. This boundary limits the feature's surface and,

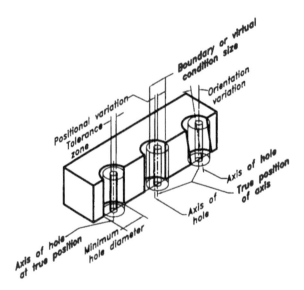

FIGURE 3-14 Equivalent controls of clearance holes by boundary and tolerance zone system.

in conjunction with the maximum material condition (MMC) concept, defines the worst case encountered in assembly operations. At the maximum material condition, the boundary identifies the greatest extent that the surface of a feature may assume (considering size and all applicable geometric tolerances). By focusing on this imaginary envelope the designer can determine the level of tolerance control required for the components comprising an assembly. As a consequence, tolerances can be based on a realistic analysis of the expected deviations. This approach is more efficient from an engineering and manufacturing standpoint than defaulting to the two-dimensional tolerances in standardized drawing notes. These default zones are not explicitly three-dimensional in nature and do not have any functional connection to the feature being considered. The apparent added effort and expense to spatially refine the product definition with geometric controls may be more than warranted by an unambiguous statement of functional requirements when considered in the context of the product's life cycle.

The second descriptive technique uses a zone containing the axis or center plane of the feature. The resulting tolerance zone is "derived geometry" having no physical reality. This is normally what is calculated and reported when inspecting a product. One predominant use for the derived zone would be in the monitoring of characteristics for process control. Derived geometry also assumes importance in the inspection process. When using CMMs, geometry is fitted using mathematical algorithms that use the point data set to derive features at various levels of abstraction (e.g., a perfect cylinder or its axis). This technical ability can be used both appropriately and inappropriately, the latter more frequently than should be expected.

Whether a boundary, an axis, or a center plane, a three-dimensional tolerance zone is provided that, when compared to conventional tolerancing, more adequately describes the variability occurring when parts enter production. In design analysis, the resulting boundary or zone provides a direct means to visualize and anticipate the cumulative effects of size and the allowable geometric controls (Figure 3-14). The controls require conscious thought and effort to apply them to a specific part but provide concrete and justifiable means to set and assess tolerances.

An example of female features (Figure 3-14) describes both the axial tolerance zone and the resulting boundary. The right-hand example in the figure shows the interaction of orientation tolerances with positional tolerances. In this case, all the allowable positional variation is used up by the lack of perpendicularity. This points out the usefulness of both methods in describing and analyzing the effects of geometric variation. The boundary technique allows the designer to visualize the effects of the tolerance controls on the resulting surfaces that comprise the component's geometry. The alternative, the axial technique, reduces the control to a derived feature (e.g., the axis of a hole) and its location within a prescribed tolerance zone. The latter provides a figure that inspectors can report.

The boundary concept is also invoked for features controlled only by size dimensions and applied conventional (plus/minus) tolerances, using what is commonly referred to as "Rule #1" of the Y14.5M standard. This is identified in the standard as the "envelope principle" (i.e., Taylor's principle) and requires a feature of perfect form if the subject feature is produced at its maximum material condition (MMC) size. As the feature departs from the MMC size, variations in form are allowed to the extent of its size departure. Control of individual features using Rule #1 does not include orientation or locational relationship between features. These additional controls require designation of a DRF and specification of the appropriate geometric tolerances.

3.2.5 Taylor's Principle

In conjunction with the boundary concept, the use of Taylor's principle, discussed briefly in the preceding paragraph, focuses the designer's attention on assembly and inspection operations. While Taylor's idea was originated with gaging operations in mind, it directly bears on the specification of geometric controls when the product and production system are designed.

Figure 3-15(a) shows a part with $\varnothing1.000-1.010$ hole. The Go plug gage, on the left end of the gage in Figure 3-15(b), must be at least the maximum thickness of the part (or depth of the hole). The gage operator may lightly rotate or realign the Go gage to start it into the hole or over the shaft being inspected. The Not Go gage, shown at the right end of the gage, need not be as long as the hole is deep. This gage should not enter the hole. The plug gages illustrated in Figure 3-15 have not been toleranced. Gage tolerances and allowances are discussed in sources such as MIL-HDBK-204A (AR), which indicates that up to

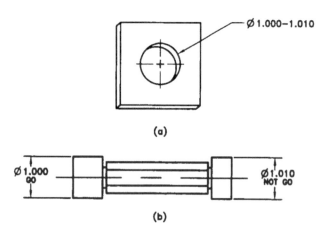

(a)

(b)

FIGURE 3-15   Part drawing and corresponding plug gage.

10% of the part tolerance, or 0.001 in this case, can be used as the gage tolerance (5% each) on both the Go and Not Go gage elements. In addition, up to 5% more of the part tolerance can be used on the Go gage as a wear allowance.

As originally conceived, Taylor's principle can be stated (King and Butler, 1957) as follows "The maximum material limits of as many related dimensions as possible or convenient should be incorporated in the Go gage; whereas the minimum material limits of these dimensions should be gaged by separate Not Go gages." Plug and ring gages that reflect the geometry of the workpiece feature they inspect do not, however, consistently accept "in-tolerance" parts. This latter statement pertains only to the Not Go gage member. Figure 3-16(a) shows the finished part hole from Figure 3-15(a) as elliptical and out of tolerance (Ø1.011). A standard Ø1.010 Not Go plug gage will not enter this hole.

Figure 3-16(b) shows the same Ø1.010 Not Go plug gage shown in Figure 3-15(b) modified so, when rotated, it may enter such an elliptical hole when the hole is too large. Figure 3-16(c) shows the complete gage. The shape of this Not Go plug gage should theoretically consist of only two points at each end of a diamond-shaped pin to completely fulfill its gaging function. A Ø0.062 dowel pin would perhaps be adequate if used in a gage as shown in Figure 3-17, providing that the shaft is parallel to the hole. If this gage could be inserted into a deep hole at a slight shaft angle and then be rocked radially, the hole would, of course,

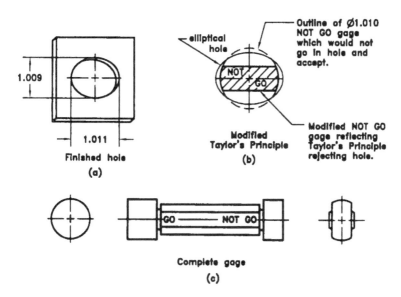

FIGURE 3-16 Part drawing and corresponding plug gage modified to reflect Taylor's principle.

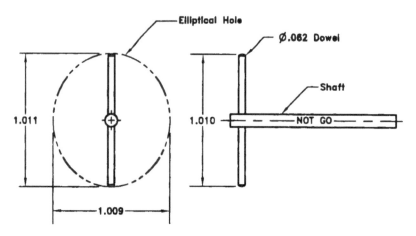

FIGURE 3-17   Pin-type Not Go gage for out-of-roundness hole (Taylor's principle).

be out of tolerance because it would be too large. A telescoping hole gage could be set at 1.010 to perform the same function; if calibrated at the beginning of a shift, it would not include either a gage tolerance or a wear allowance.

Figure 3-18 shows Taylor's principle applied to the Not Go gage for a shaft. Fortunately, most snap gages reflect this principle.

Many part features shown in this text require plug or ring gages to check their minimum and maximum sizes as the first level of control. The Go plug or ring gage is not required, however, if a locational or orientation tolerance of zero at MMC is specified, since the functional gage will include the Go gage element. Separate Go plug or ring gages are never needed to ensure functional interchangeability in cases where fixed-element functional gages simulate the mating part at its virtual condition. The key word in this last statement is "interchangeability" Also note that it incorporates the term "virtual condition" rather than the word "size."

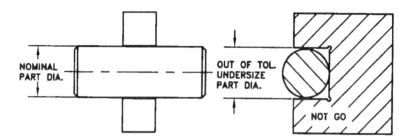

FIGURE 3-18   Application of Taylor's principle to ring gage.

The virtual condition concept (as described in Y14.5M) can expand Taylor's idea. The virtual condition is the boundary generated by the collective effects of all applicable geometric controls and the MMC size of the feature. These effects enlarge the boundary of an external feature or reduce the boundary of an internal feature.

The conceptual effect of expanding Taylor's principle requires simultaneously gaging all virtual condition limits of all functionally related features on a part. As this simultaneous gaging is performed, product assembleability is assessed. The intent is to include all functional features related by the design. Such gaging applications require specification of the fundamental DRF, which Taylor's gages did not identify. At the other extreme, the least material condition of these features is not likely to cause assembly problems and may be inspected using individual Not Go gages. The least material condition is not going to impair assembly under normal circumstances but could lead to structural failure if not verified.

By combining the envelope concept and the datum reference frame, the design team is given a realistic method of discussing the interrelationship of actual part features. Application of these concepts assumes the role previously played by the physical prototype. While not replacing such a prototype, it does extend the design process into areas that conventional dimensioning and tolerancing techniques do not, or cannot, consider.

3.2.6 Refinement of Controls

All manufacturing endeavors have economic constraints. This is true even in the case where a single unit of product is made. The more sophisticated techniques found in the Y14.5M standard are applied in any situation where the additional refinement of size and conventional tolerances is necessary to yield the desired functional results. The refined definition provided by adding controls does not automatically increase product costs.

In one sense, the first level of control invoked deals with the workmanship standards found in a firm's shop. If these are sufficient to achieve the desired results, then no further controls may be desired and there will be an absence of documented relationships and controls on the drawings. A supplemental tolerance block on the product drawings can be added to provide generic tolerances for dimensions. The authors do not recommend these techniques; however, they recognize it is a feasible business decision and is likely to be encountered. Where the product is designed and produced at the same facility, it may be the preferred method.

In the event that the default tolerances do not provide the necessary control, then conventional, plus/minus tolerances should be applied directly to the individual dimensions where required. The interaction of this size control with Rule

#1, provided Y14.5M is invoked on the drawing, may provide sufficient control to maintain component functionality. It is again emphasized that the application of Rule #1 neither controls the relationship between features nor gives complete three-dimensional control. As a consequence, a conventionally toleranced component detail does not provide a complete product definition. Application of conventional tolerancing assumes that the design is sufficiently robust to be unaffected by lack of complete definition. Actual shop tolerances and machinist skill in interpreting the drawings are presumed to give the desired functional relationships.

It should be reiterated that these first two levels of control yield their best results when highly skilled and experienced personnel do all the work in-house. In these circumstances, the designer can make a quick trip to the production area to address any problems concerning interpretation of the design. As the work begins to move out of the company's facilities to vendor's plants, it requires more definitive specifications (particularly tolerances) and motivated personnel to resolve the problems. A vendor cannot produce an acceptable product from an incomplete product definition.

Where the above techniques do not achieve the desired function or relationship and the economics of the design warrant their use, geometric controls should be added to the product specification. One of the advantages in using geometric controls is that they require a conscious effort to define the type and degree of control to be applied well before the design is released for production. Thought precedes the expenditure of resources. When applying these controls, the resource most likely to be conserved is that which is most precious—time. There is no dispute that, given sufficient time, an acceptable product definition can be achieved by trial-and-error. The problem is that each trial adds to the cost and consumes time.

Applying any category of control (from size control through GD&T and surface texture) involves decisions about the level of control needed. The designer should be led sequentially through the levels of refinement, making conscious decisions about the effects of the proposed controls on function and producibility. When these decisions dictate the use of geometric controls, the reference frame concept and geometric control provide a checklist that can be used to lead even the novice designer through the steps necessary to arrive at a functional product definition.

3.3 APPLICATION CONSIDERATIONS

Once introduced to the general concepts, a designer's success in adequately defining a product's geometry requires decisions recognizing the reality of production. In general, this reality deals with the physical variation inherent in production and inspection and the challenge of controlling it.

3.3.1 Production Variation

The standard provides a differentiation between the designer's vision of the perfect product and the results of the production. This is most directly seen in the variety of terms to deal with the concept of the datum (e.g., datum, datum feature, and datum feature simulator), all of which have concise definitions. Geometric control emphasizes the variation inherent in producing interchangeable components and forces consideration of this variation in all aspects of product design and manufacturing. To achieve sufficient control of the process, the design team must use a common language to discuss the type of variation that can occur and to subsequently restrict its magnitude. Production variation must be partitioned into identifiable and controllable elements for realistic process control.

3.3.2 Datum Accuracy

Since all the features found on a component are defined with respect to its location in space, one of the major sources of variation is associated with the datum reference frame. These variations can be placed in two major categories.

The first relates to the individual datum features. The quality required of these individual features is determined by the functional design requirements. By looking at the tolerances on controlled features that reference the DRF, the level of accuracy of the datum features can be established, with accuracy ratios of 4 to 1 or 5 to 1 to 10 to 1 being common. In the absence of other technical methods to determine the level of control of the datum features, the specification of datum accuracy should be based on an analysis that considers both the total process variation and the variation introduced by the inspection system. The necessary level of datum accuracy ultimately depends on the ability to verify—including a statement of the uncertainty in the measurement—the specified control of the component's features. An appropriate specification results from a stable datum reference frame, allowing repeatable and reproducible part measurement.

The second major item reflects concerns about datum accuracy during processing, measurement, and assembly. The standard makes distinctions among (1) the datum designated by features in the product description, (2) the datum features on actual parts, and (3) the datum feature simulators that are part of the processing equipment. The latter items include the table of a machine tool, a special fixture, or a CMM surface plate. Techniques used to arrive at the needed level of control focus on the variation inherent in the production process and allied equipment. Recognition of this process variation must be included in the design cycle.

It should be highlighted, based on typical industrial practices, that a large proportion of the variability inherent in processing and measurement can be traced to variation in the setup. Incomplete DRFs or the use of part features not identified as datums are a significant source of setup error.

3.3.3   Interchangeability and Assembleability

While the techniques this text offers may be useful in creating any type of product, the greatest concentration of applications is for products requiring complete, random interchangeability. As implied in the discussion of Taylor's principle, the overriding concern in being able to assess the level of interchangeability and assembleability is the quality and relationship of the functional features that provide assembly location.

Specification of a DRF allows a direct and concise description of the required relationships. Without this level of design refinement (i.e., the reference frame), there is no consistent communication of the relationships needed to successfully assemble a functional product under the conditions imposed by state-of-the-art manufacturing. Experienced designers and process planners recognize that the choice of a process DRF rather than the functional DRF introduces tolerance stacks; these uncertainties may defeat the best design unless consideration is given to analyzing and controlling their effects prior to production.

3.3.4   Verifiable Controls

If a design contains geometric controls, the designer is responsible for relating these controls to a specific product function and for specifying verifiable controls. It serves no purpose to ask for types or levels of control that cannot be verified or that do not verify the intended function. As a corollary to this, unverifiable controls may be nonfunctional controls. If no method exists to verify the callout on the drawing, then the mating part may not actually see the effects the designer anticipated.

An example of nonfunctional verification would be the prevalence of two-point measurements to inspect the size of a feature that must assemble with a mating part. As Taylor's principle clearly points out, such measurements provide no information relating to assembleability. If the desire is to learn something about the impending assembly process, then a functional (full-bodied) gage should be used that simultaneously inspects all the features related to the part's assembly. If a pin is to enter a bore, the inspection of pin size using a micrometer (a two-point measurement) will not ensure that the pin is sufficiently straight to allow assembly.

The Y14.5M standard provides a list of allowable controls with concise definitions—mathematical precision is added by the Y14.5.1 standard (ASME 1994b). It is a written source that can be used to arbitrate disputes related to the controls placed in the product definition. While the standard is not intended to define possible gaging techniques for a particular control, it provides enough guidance to allow intelligent definition of the methods that could verify conformance. This should allow the concurrent engineering team to reach a consensus.

The ultimate and most realistic arbiter of whether a control is verifiable is if the part can be set up using surface-plate or CMM measurement. With the possible exception of measuring circularity and cylindricity using a surface plate, a preliminary check on the controls of average-size parts can be instituted by these relatively inexpensive methods without resorting to fabrication of hard gaging. If the part cannot be set up for measurement in either of these fashions, then the drawing control is not understood or an inappropriate control has been specified.

3.3.5 Material Modifiers

Significant design and tolerance effects are introduced by placing a material modifier on a feature's geometric control. Of particular concern is the effect on the design of tooling and gaging. The indiscriminate use of an inappropriate modifier can add considerable cost to the product and yield components that do not provide the required function. This is the consequence of tool and gage designers' altering design intent for the convenience of the tooling. Examples that relate to the callout's impact on gaging (and tooling) follow.

Regardless of Feature Size (RFS).   Tolerances modified with the RFS (regardless of feature size) callout (Figure 3-19) are *constants* and cannot be increased. The RFS specification is not a practical gaging callout because a series

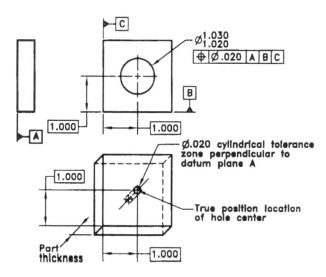

FIGURE 3-19   Constant diameter cylindrical tolerance zone.

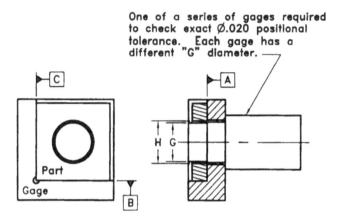

FIGURE 3-20   Gage cited in Tables 3-1 and 3-2.

of gages, all with different, interchangeable diameter pins (dimension G in Figure 3-20), would have to be purchased to check the parts when the finished holes varied in diameter between 1.020 and 1.030. Table 3-1 shows the different gage pins required. However, even a series of gages, as shown in the table, would not exactly meet the RFS specification unless they were exactly 0.020 smaller than the finished workpiece hole size. The RFS tolerance modifier would therefore actually require gage sizes as shown in Table 3-2. Thus, to gage this part with its RFS callout, an infinite range of gage elements—an impossible requirement—would be needed. A possible solution to this situation would necessitate a large number of gages in 0.0001 increments, which is still not practical. The RFS callout does not specify the use of fixed-size gaging elements, which functionally represent the strict interchangeability requirements of most mating parts.

TABLE 3-1   RFS Gage Element Diameters

Finished hole dia. H (in.)	Gage element dia. G (in.)	RFS positional and perpendicularity tolerance (in.) (a nonvariable tolerance)
1.020	1.000	0.020
1.021	1.001	0.020
1.022	1.002	0.020
1.023	1.003	0.020
1.024	1.004	0.020
1.030	1.010	0.020

TABLE 3-2   Effect of RFS Modifier on Gage Sizes

Finished hole dia. H (in.)	Gage element dia. G (in.)	RFS positional and perpendicularity tolerance (in.) (a nonvariable tolerance)
1.023	1.003	0.020
1.0327	1.0127	0.0200
1.02981	1.00981	0.02000
etc.	etc.	etc.

Maximum Material Condition (MMC).   Tolerances modified with the MMC callout are variable tolerances and may increase as the part feature is finished away from its most critical interchangeability size. This specification is a practical gaging callout because a gaging element of fixed size and location will automatically allow part hole location and perpendicularity to vary as finished holes vary in size. This fixed dimension is the size of the virtual condition boundary. Table 3-3 shows that a fixed-size gage element (Figure 3-21) used to gage the finished part shown in Figure 3-22(a) will allow varying tolerances. This fixed-gage-pin element could represent a strict design requirement: For instance, a ∅1.000 bolt must pass through the hole.

If Figure 3-19 had specified MMC instead of RFS [see Figure 3-22(a)], the positional tolerance could vary, as shown in Table 3-3. Note that since the positional tolerance specified (∅0.020) must be held only when the hole is at MMC (∅1.020), this tolerance may increase by the exact amount the hole departs from its most critical (MMC) size. The MMC specification, if substituted for the RFS callout shown in Figure 3-19, requires the ∅1.000-gage pin shown in Table 3-3 and illustrated in Figure 3-21. This ∅1.000-gage element is determined by subtracting the ∅0.020 true-position tolerance zone specified at MMC from the MMC size of the hole; thus,

TABLE 3-3   Positional Tolerances at MMC

Finished hole dia. H (in.)	Gage element dia. G (in.)	RFS positional and perpendicularity tolerance (in.) (a nonvariable tolerance)
1.020	1.000	0.0200
1.0215	1.000	0.0215
1.0260	1.000	0.0260
1.0272	1.000	0.0272
1.0291	1.000	0.0291

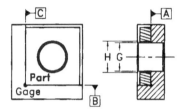

FIGURE 3-21   Gage cited in Table 3-3.

1.020 MMC or most critical hole size

− 0.020 true-position tolerance diameter allowed at MMC

1.000 diameter fixed-gage feature

The following general rules should be followed in the use of MMC gages:

1.   The gage pin should be the same form as the part feature and the true-position tolerance zone
2a. Subtract the positional tolerance from the MMC size of internal features to obtain the basic gage element.

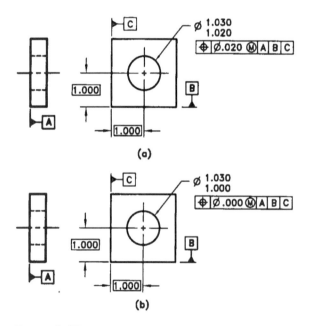

FIGURE 3-22   Conventional and zero tolerancing.

2b. Alternatively, add the positional tolerance to the MMC size of external features to obtain the basic gage element size.

Referring to Figure 3-21, dimension H minus dimension G equals a variable
tolerance when MMC is specified. The true functional requirement always remains fixed and is represented by gage pin G. Due to inherent process variations,
parts will vary and no two part features will ever be exactly alike.

Since the $\varnothing$1.000-gage pin checks only the hole location and perpendicularity at MMC, the size of the hole must be checked separately with $\varnothing$1.020 Go
and $\varnothing$1.030 Not Go gages. However, since the $\varnothing$1.000 fixed-gage pin simulates
a $\varnothing$1.000 bolt in the mating part, the use of a separate $\varnothing$1.020 Go gage will
reject the part if the hole is smaller than 1.020—even if this hole is accepted by
the $\varnothing$1.000 fixed-gage pin. Thus, the very best parts (with holes undersized but
located more precisely than the $\varnothing$.020 positional tolerance specified) will be
rejected.

Millions of dollars are wasted each year because of the traditional use of
such Go gages. However, as previously shown, there is a way out of this dilemma.
The separate $\varnothing$1.020 Go gage can be eliminated and incorporated in the functional gage if the specifications shown in Figure 3-22(b) are used. A callout that
requires a "perfect" hole ($\varnothing$.000 position error) at MMC makes the positional
tolerance *completely dependent* on hole size, as shown in Table 3-4. The callout
saves separate Go gage cost and operator time and also allows the manufacturer
the freedom to pick his or her own working tolerance when he or she chooses
a drill size.

3.3.6 Basic Interchangeability Gages

To illustrate the application of the preceding ideas to assembly gaging, the following gages will guarantee the interchangeability of the parts in cases I, IIA, and
IIB. Each part feature (hole, tapped hole, etc.) in each pattern of features has the
same basic locations.

TABLE 3-4 Relation Between Positional Tolerance and
Hole Size

Finished workpiece hole dia. (in.)	Positional and perpendicularity tolerance dia. (in.)
1.001	0.001
1.007	0.007
1.017	0.017
1.021	0.021

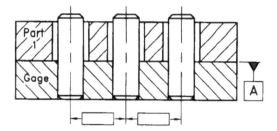

FIGURE 3-23 Gage for part 1 (clearance holes).

Case I: Clearance holes in mating parts
Rule: The gage for each part consists of a pattern of pins located at part basic hole locations. The gage pins will be the MMC size of the assembly screws or bolts (Figure 3-23).
Case IIA: Clearance holes in part 1 (Figure 3-23) and dowel pins or studs in part 2 (Figure 3-24)
Rules: *Part 1*
The gage for part 1 (Figure 3-23) consists of a pattern of pins located at part basic hole locations. The gage pins will be the MMC size of the part 2 dowel or stud plus the positional tolerance diameter specified at MMC for the dowels or studs. The positional tolerance increases the virtual size of dowels or studs, and makes them effectively larger at assembly.
Part 2
The gage for part 2 (Figure 3-24) consists of a pattern of bushings located at part basic dowel or stud locations. The bushing IDs will be the part 2 dowel or stud MMC size plus the positional tolerance diameter specified at MMC for the dowels or studs.

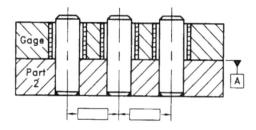

FIGURE 3-24 Gage for part 2 (studs or dowels).

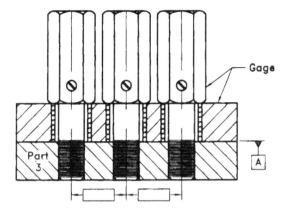

FIGURE **3-25**   Gage for part 3 (tapped holes).

Case IIB: Clearance hole in part 1 (Figure 3-23)
            Tapped holes in part 3 (Figure 3-25)
   Rules: *Part 1*
            The gage for part 1 consists of a pattern of pins located at part
            basic hole locations. The gage pins will be the part 3 tapped
            thread size plus the positional tolerance diameter specified at
            MMC for the tapped features.
            Part 3
            The gage for part 3 (Figure 3-25) consists of a pattern of bush-
            ings located at part basic tapped-hole locations and a series of
            Go thread gages, one for each tapped hole in the pattern. The
            difference between the bushing ID and the shank diameter of
            the Go thread diameter where it goes through the gage bushing
            will be the positional tolerance specified at MMC for the
            tapped thread. The bushing length will equal the thickness of
            the mating part.

3.4   SUMMARY

To achieve acceptable results, product designers must identify the variation inherent in all manufactured articles, assess the magnitude of this variation, and develop methods to control it. The greatest success in accomplishing this challenging task occurs within a system of structured design, one that integrates key
organizational elements with the technical design system. Design techniques built
on this foundation provide a technical system that most effectively operates
within the context of team-based concurrent engineering.

The key concept contained in this chapter involves the datum reference frame. This framework provides the base on which the product and its constituent components are built. It is the focal point for all the geometry constituting the product and provides a common thread woven throughout the design process. The reference frame establishes the origin for the measurements associated with component features. As a result, all the design and process decisions ultimately rest on the appropriate choice of a DRF.

These design techniques are predicated on a product that is both interchangeable and producible. Taylor's principle is a necessary element in achieving these results. By expanding Taylor's intent to include the virtual condition concept from ASME Y14.5M, the necessary geometric relationships are identified, leading directly to a design definition compatible with the needs of competitive manufacturing.

With the variety of geometric controls available, the virtual condition boundary concept provides the means to select and evaluate the appropriate controls. Additional importance is attached to the boundary because it allows designers to assess both the interchangeability and the manufacturability of assembled components, ultimately providing truly verifiable controls.

All controls should be verifiable. The uncertainty of the methods used to measure the product should be of reasonable magnitude (generally not consuming more than 10% of the part tolerance), allowing the results of the measurement process to be meaningful. In conjunction with this, the various controls and material modifiers must be carefully considered during concurrent engineering for their effect on the design and cost of functional tooling and gaging.

A reasonable understanding of the place that each of the concepts holds in a system of structured design provides the first step in mastering geometric control. Each forms part of the framework that students of the system can use to provide order to the detailed applications encountered in product development. Such a framework is needed because each control application is unique, depending as it does on the functional requirements of a specific design and precluding memorizing examples to cover every possible case.

REFERENCES

ASME, Y14.5M-1994, Dimensioning and Tolerancing, New York: ASME, 1994a.

ASME. Y14.5.1M-1994, Mathematical Definition of Dimensioning and Tolerancing Principles, New York: ASME, 1994b.

King, G. K. and Butler, C. T., Principles of Engineering Inspection, London: Cleaver-Hume Press, 1957.

4

Design Layout

4.1 INTRODUCTION

The preceding chapters have set the stage for the real work of product and process design. The results of these efforts will be seen in the completed definition (e.g., drawings, process plans, tool and gage designs, work methods, and more) needed to actually manufacture the product. Unlike many professions, manufacturing provides a tangible, physical result that can be realistically evaluated.

The following chapter presents a logical process to be used in defining the product at the system level once the design concept is chosen. This material touches on creative design only to demonstrate how the original concepts are systematically developed; then it presents a methodology to be used in defining and then refining the geometric form of the product assembly. Keeping with this book's focus, the material shows how the product and process are rationally moved from concept to detailed design. The results of this work will have a tremendous impact on the product's cost structure and, ultimately, on its success.

4.2 PRODUCT ARCHITECTURE

A variety of techniques may assist in determining product architecture. Some of these are unstructured, relying on serendipity for the plan to come together; others are highly structured, forming the elements of a sequential process that gives definition to the design. Two examples tied to specific design tools show how these techniques can add value to the product cycle.

As the company, including suppliers and customers, becomes more knowledgeable about product design, one technique that can be used in creating a complex product with a host of characteristics is *quality function deployment* (QFD). This technique (King, 1989) allows what is commonly referred to as "the voice of the customer" to enter the process in a way that ensures that the design is focused on specific market needs. In a more technical vein, QFD translates the customer's desires into a form the design engineer can use. A formal process such as this seeks to translate customer requirements into specific goals to be achieved by the design and to provide quantitative measures to guide the process.

A short anecdote illustrates the importance of this step. One of the authors once worked for a firm that undertook the redesign of a major portion of its industrial product line. The company's initial intent—not necessarily the customers'—was to improve the design of the equipment while removing cost from the manufacturing process. The driving force behind doing this was increased market pressure from foreign competitors. The results of the project were horrible. It took longer than expected; it consumed more money than was budgeted; and the redesigned product cost more than the previous generation. What it turned out to be, admittedly in hindsight, was a engineering exercise that was never driven by market needs. The engineers forgot who the customers were and what they wanted.

A second example further demonstrates the need to translate a customer's desires into precise numerical values to be used in setting engineering specifications. An electronics manufacturer was developing a switch based on preliminary drawings included as part of the customer's request-for-proposal. During discussions of the design, one of the marketing people pointed out that the existing switch provided by the competition generated an audible click when actuated. The customer's engineers had indicated that this was needed in any switch that they would likely buy. During subsequent discussions, essentially a preliminary design review, this new characteristic was treated as one of the items in the specification. However, it was not formally included on the customer's drawings and there was no means (a metric) determined that would allow a unique specification to be created. So a very subjective element was informally added to the design, an element that might derail subsequent work.

The results of QFD describe the functional requirements and provide metrics used in design, manufacture, and assessment. At some point in the process, the various subsystems have to be identified and specific functions allocated to them. Another tool, *structured requirements analysis* (Grady, 1993), is one of the methods developed to do this. Created within the defense systems industry, it takes the mission of the system and breaks it down into elements that can be used in designing subsystems and components. Some aspects of *value analysis* (Fowler, 1990) provide yet another method to accomplish a similar task. The goal in applying any one of this group of tools is to provide a functional definition of the product architecture. This in turn provides the criteria used in giving geometric definition to the product. The common thread found in these techniques is a verbal model focusing on the "how–why" relationships of the product. This identifies the functional relationships of the envisioned product before it takes physical form. It directs the activities toward designing the product in toto rather than designing individual components (or subsystems) with the assumption that the desired product (i.e., the system) will emerge from the individual component designs.

The interest at this stage of the design cycle—and of this chapter—is in creating a baseline system design, specifically the design layout. The design team then proceeds with the detailed design process, reducing the concept to components that can be economically produced. However, the design layout cannot be used to accomplish this task if the function of the various subsystems and components is not well defined; hence the need to apply QFD (or some equivalent method) to define the product architecture. No argument is made to specifically use QFD or other structured techniques. However, without the conscious use of something similar to these processes, it is likely that an implicit definition is put into place through happenstance. When the cost accountants finally get to do their work, the results may be very unsatisfactory. It is more efficient to avert surprises by using techniques that provide structure and focus to a project. The industrial product redesign mentioned is an example of such unexpected and unwarranted outcomes.

The techniques that should naturally follow the definition of a detailed architecture (see Figure 4-1) are developed in the next section. This architecture

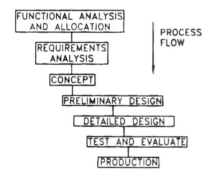

FIGURE 4-1   Design cycle.

specifically includes the functional requirements allocated to each of the subsystems. The six-step methodology discussed next begins the process of giving these functional elements geometric reality.

4.3 THE SIX-STEP METHODOLOGY

The concurrent engineering team follows a series of six steps (Figure 4-2) to achieve the required product and process system design. While these steps may appear to be tedious and time-consuming, all the information that flows from the methodology must be generated somewhere within the company before the product and process design is released. It can be done in a logical and controlled fashion, or it can be done haphazardly. As the complexity of the product increases, the information resulting from these steps will eventually be developed by the conscious use of the methodology or through trial-and-error. In every case, in reaching the same goal (i.e., a customer-needed, deliverable product) either method must eventually amass the same knowledge base. The importance of our six-step methodology is that it develops this information at the appropriate design stage where it is the most economically beneficial. If, for instance, the tolerance analysis is performed after the tooling stage of the production cycle where physical elements of the system exist, then any necessary changes will be more expensive than if they occurred during a design stage where there are no physical assets. The steps are performed in a logical order to conserve precious resources.

The team does not need to strictly adhere to the methodology; it can be modified to reflect the nature of the product and the strategic goals of the firm. Not all of the steps need to be formally implemented on every project. Decisions based on experience may replace the rigid use of the discrete steps that will be outlined. However, shortening the process in this manner should be done with

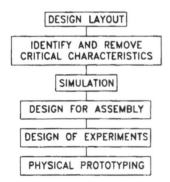

FIGURE **4-2**   Six-step methodology.

the consensus of the concurrent engineering team. Due to interactions within the design cycle and among functional groups, only the team should decide to depart from any of these specific procedures.

4.3.1 Design Layout

Once the conceptual phase of the product design is complete, a preliminary design layout (Figure 4-3) of the new product should be performed. This layout is used as a baseline for discussions and decisions as the design work progresses. It is a more detailed representation of the concept that has survived the first design phase. It is no longer in the form of concept sketches used to assess the feasible design alternatives. The layout contains sufficient shape and size information to allow informed discussions of the proposed geometry. All subsequent activity is based on this layout, and the production version of the product will be derived from it. The method used to develop this information is crucial to the efficiency of the design project.

Three general approaches to product design exist. They can roughly be characterized as (1) the *design layout*, which is seldom used, (2) the *assembly drawing*, which is sometimes called a *zero drawing,* and (3) the "*quick and dirty*" preparation of detail drawings with no layouts preceding them.

The design layout is a unique assembly drawing. It geometrically defines the nominal size and the limits of size and location of all parts that make up the assembly. The most critical limits are usually shown in green (the MMC or Go gage size), the least critical limits in red (the Not Go gage size), and the nominal

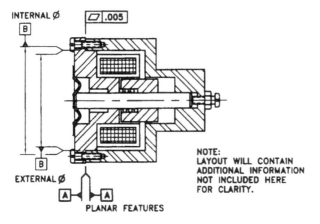

FIGURE 4-3 Partial design layout for pump assembly.

in black. The design layout is made at the level of complexity needed to aid analysis. It allows the designer to quantitatively determine the effects on part function when all components are at their critical functional limits of size and location. In many cases, overlays or CAD layers will prove useful when looking at these relationships. The design layout enables the designer to utilize the largest tolerances of location and the broadest limits of size, including form and orientation effects, possible before function is impaired. At a minimum, the following technical information should be contained in the preliminary design layout:

Functional datum identification
Qualification of datum features (accuracy ratio)
Critical dimensions and relationships based on classification of characteristics
Interchangeability parameters
  Functional limits of size and location
  Fits and allowances
Special materials and processes
Standard parts

The design layout serves to

Enforce preparation of sketches at the nominal conditions along with the appropriate boundary extremes (virtual and resultant conditions) so that the design can be functionally modeled in three dimensions
Allow preparation of all details, support drawings, and acceptance procedures
Allow structural and stress calculations for the appropriate boundaries
Support the design of tooling including jigs, fixtures, and gages
Enable life testing and material capability analysis
Determine the design's manufacturability

All members of the concurrent engineering team should review the first attempt at creating this design layout. Items to be specifically evaluated include functionality and accessibility of datum features, the accuracy ratio needed to provide the necessary control between the DRF and the controlled features, justifiable tolerances, and critical relationships between features that must be included in the product definition.

The major value of the design layout is that it forces the team into early consideration of design simplification when the cost impact of change is less significant. Changes that occur after detail drawings have been completed, or later after the design has been released, become increasingly costly. Changes that occur after production has started are the most costly and unacceptable of all, in terms of both time and expense. In fact, these changes may be impossible to achieve, and the company may be forced to produce an unprofitable product.

4.3.2   Identification and Removal of Critical Characteristics

The characteristics of the assembly that have the greatest influence on product function are identified next. One useful method of classifying these characteristics is found in DOD-STD-2101, where the categories are critical, major, and minor. The concern at this stage is with the critical and major classifications. Adapting definitions of these terms to use in the commercial sector produces the following: Critical characteristics involve human safety and certain product failure; major characteristics may cause failure of the product function. Minor characteristics are not likely to impair product performance.

Failure mode and effects analysis (FMEA) provides a formalized application of the idea inherent in these definitions. This technique (Stamatis, 1995) identifies possible failure modes in the proposed design and is used to prevent a product that exhibits such failures from getting to the customer. Failure-proofing of the design should take place by eliminating as many of the critical and major characteristics as possible. Of particular interest in this book is the use of FMEA in conjunction with the reference frames identified on the baseline layout. Since the datum features that comprise the DRF are the crucial elements in placing the component into its functional position, the geometric characteristics of each datum feature should be analyzed to determine how it might affect product failure. In the event that the FMEA identifies any of these features as a factor in a possible failure, the design team would try to redesign the product to remove the critical element.

Similar procedures should be applied to the FMEA for all critical and major functional features—distinct from datum features. With this discipline included in the design cycle, the worst outcome would be the inability to remove the offending feature while being able to refine the inspection process to prevent imminent product failures from reaching the customer. Yield might be affected, but the customer would see no defects.

It is at this stage that the concurrent engineering team begins to have its greatest influence. Because removing these characteristics puts the design in a state of flux, it is obvious that members of design engineering, marketing, manufacturing engineering, and the other interested disciplines need to be actively involved. These decisions are not to be implemented until the team reviews them.

To illustrate the value of this step, a solenoid pump is shown in Figure 4-4. A complete discussion of the design is reserved for the next chapter. The team's interest is in simplifying the design to eliminate any possible failure modes

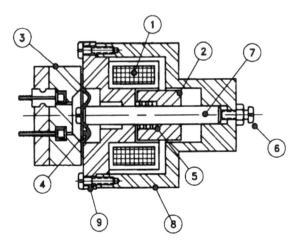

FIGURE 4-4   Pump assembly.

and, possibly, improve its manufacturability. During this stage of product development, the design contains at least one geometric problem that might be eliminated by a design alteration. The rod, identified as number 7 in the figure, is located within two bores that are features in two separate components. From a functional standpoint, the positions of each bore must be related to the other such that the rod can be actuated without binding. The actual position of each bore is controlled by separate DRFs that are contained on the pump body and on the endcap, respectively. The bore in the body is located and controlled with respect to the DRF established on the body (see Figure 4-5). The bore on the cap is positioned (located and controlled) by the DRF established on the endcap (Figure 4-6). However, when assembled, the bore on the endcap is positioned by a combination of its location within the endcap's DRF and the interrelationship of the cap's datum features with the mating features on the pump body. These serial influences (Figure 4-7) determine the variation exhibited by the assembled components that, in turn, determines their ability to function. In this case, misalignment of the bores could cause the rod to bind, reducing or preventing pump action.

This example shows how DRFs and geometric controls may be used to produce a list of critical and major characteristics. For the pump, the two separate DRFs that position the bores can cause greater variation, a tolerance stackup, than would occur if the two bores were positioned within a single DRF. By critically reviewing the design layout for information concerning the DRFs and features related to them, failure modes influenced by the geometric characteristics can be identified. In this case, the concurrent engineering team redesigned the solenoid

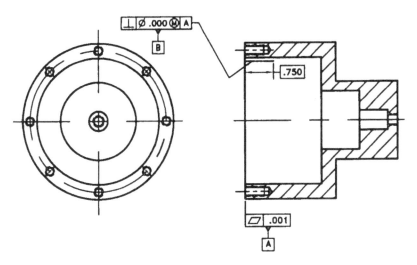

FIGURE 4-5 Detail for pump body showing datum reference frame.

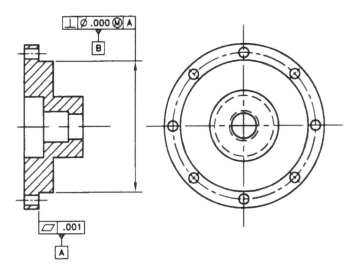

FIGURE 4-6 Detail for endcap showing datum reference frame.

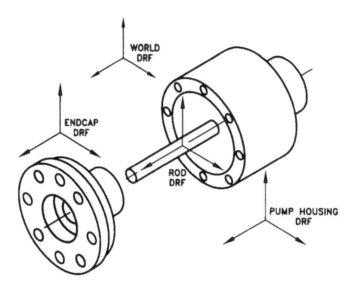

FIGURE 4-7   Illustration of chained DRFs.

pump to combine the effect of the two bores in a single component (Figure 4-8). Both features are now contained within a single datum framework that minimizes the variation the pump assembly will experience.

The above example deals with a design-critical characteristic. There may also be process-critical characteristics that can be eliminated by redesign. Referring to Figure 4-8 again, the cap that is now part of the design does not contain the more complicated geometry of the original cap design, which provided one of the datum features in the form of the turned diameter. The removal of this

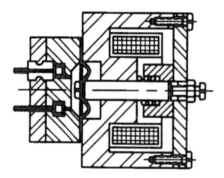

FIGURE 4-8   Redesign to remove critical characteristics.

diameter, a design-critical feature, has also allowed simplification of a process-critical feature and improved manufacturability.

4.3.3 Mechanical Simulation

The next step creates a complete prototype embodying the effects of process variation. The authors' experience has shown that more than one prototype may be proposed at the beginning of this stage. Each disciplinary representative to the team may suggest a version that enhances his or her functional interests. Manufacturing may come to the table with a design that enhances productivity, while design engineering may produce a purely functional design that does not consider producibility. While a physical prototype is probably the most desirable method to simulate the design, the number of trials necessary to resolve design issues and optimize the design with such a prototype is usually unacceptable because of the time involved. Additionally, the expense incurred in executing numerous physical versions of the baseline prototype design would be prohibitive. The increasing use of computer-aided techniques to simulate the product and aid in design optimization provides a cost-effective solution leading to a prototype acceptable to all team members.

Computer techniques now make it possible to simulate static, dynamic, kinematic, and spatial relationships and achieve many of the results that were previously reached only through the use of physical prototypes. It is also possible to model some elements of the production process with the use of computer application packages. One prominent example is the mold flow analysis that can be routinely performed in the design of injection molded components. The effect is to reach a level of product definition that, in the past, would have been gained only through the expenditure of much time and money; in fact, it may have been reached only after the product was actually in production. The use of these simulation techniques can now anticipate problem areas that, in the past, were discovered only after the system was fully implemented.

This stage of the product design, as mentioned, should involve the simulation of the specific spatial relationships needed to ensure product assembleability and function. The design team's use of the ASME Y14.5M standard is of critical importance. In simulating spatial relationships, geometric dimensioning and tolerancing has a threefold purpose: It provides a common technique to identify the functional requirements placed on the geometry; it communicates these desires in precise and accurate form; and it allows the effects of the geometric controls to be incorporated within the simulations.

The major advantage in using standardized geometric controls in the design is that the underlying language describes anticipated deviations in terms common to the manufacturing disciplines. The resulting baseline model will be more efficient (i.e., will contain more useful information) and yield more realistic results if the simulation can anticipate the downstream activities of the manufacturing process. The concurrent engineering team can most effectively set design parameters and tolerances for the simulation if well-defined variations are built into the model; hence, the use of the Y14.5M standard.

As discussed earlier, the primary intent of the creative design phase is to provide a definition of product function without unduly restricting the ultimate geometric design. This preliminary stage does not generate a robust product definition that can immediately be placed into successful production. The GD&T controls are brought to bear on the idealized design and refine it to a point that becomes achievable in the physical world. The team uses these controls to guide them when considering product deviations from perfect form, orientation, and location that can affect functional relationships. At this point in the cycle, these variations have usually been either ignored or given insufficient thought. Use of the controls slows down the design process at just the appropriate point to take a critical look at the idealized product definition. While this technique lengthens the simulation phase, products exhibiting characteristics that require the use of geometric controls will consume time during this phase or later during the numerous engineering change cycles. This latter method of ensuring control is much more expensive and does little to build confidence in a firm's design abilities.

4.3.4   Consideration of DFA

The team should now review the design using the appropriate "design for . . . " techniques. Various sources (Boothroyd et al., 1982) show that the cost of assembly in many products can exceed 50% of the total manufacturing costs. Due to the large impact of assembly cost and other competitive concerns, product assembly should be investigated from the standpoint of both design and processing. One of the more dominant techniques in this area is *design for assembly* (DFA).

While whole texts have been written on this one technique, what is of concern here is the requirements that DFA imposes on the overall design of both product and process. Specifically, a number of general design rules attributed to Boothroyd (1982) facilitate both manual and automatic assembly. The geometric controls from Y14.5M are necessary to implement these guidelines.

The first of the guidelines directs the designer to minimize the number of parts in the product. This admonition is entirely compatible with the elimination of critical and major characteristics undertaken in an earlier step of the methodology. As the number of components is reduced, the density of functional part features contained on a single component is likely to increase. As such relationships multiply, it becomes exceedingly important that these relationships are identified and unambiguously communicated. The only consistent method to describe the desired functional relationships is through the use of the DRF and geometric controls.

An example using geometric controls illustrates the need to consider assembleability at this point. If the density of functional features on a component increases as a result of a DFA process, it is likely that many of the controls called for will be positional controls that describe the correct mating relationships. Where all of these controlled features must be assembled simultaneously [see Figure 4-9(a)], using a positional control with the same DRF and modifiers communicates the desired functional requirement. Where separate assembly requirements are needed, adding the phrase "separate requirements" as shown in Figure 4-9(b) can break the linkage between features. This second situation relaxes inspection requirements and may increase process yield. The geometric dimensioning and tolerancing standard provides the only method that incorporates assembly requirements into a robust product definition and that ensures complete understanding of design intent.

Another DFA technique illustrating the importance of geometric controls is the use of a base part to provide an assembly platform to receive other components. This is an excellent example of the need to identify and maintain functional relationships. In our geometric language, the base is a unique DRF that controls the desired relationships as features or components are added to the assembly. In particular, Boothroyd mentions that this base part should be "readily located in a stable position." This is the purpose of the DRF; locating the part in a singular position in three-dimensional space. Conventional methods of part description either imply the DRF or do not specify it at all. When dealing with individual components and the parent assembly, the desired kinematic control can be precisely stated only by specifying the DRF.

Whereas these two principles deal with the complete or assembled product, other principles apply to individual components. The main guidance concerns the need to orient these parts in the appropriate position for assembly, with the base part serving as an assembly fixture. The concept of the DRF again provides the mechanism to describe the part's position (geometric function) in the completed assembly and aids in subsequent design of tooling and gaging. Furthermore, the DRF is the first step in the process of developing geometric controls that anticipate component variation, improving both producibility and assembleability.

4.3.5 Design of Experiments

The intention of this phase of the cycle is to ensure the product's functional integrity. The target parameters that in combination determine the product's functional response are established and their sensitivity to variation analyzed. By using statistical design of experiments, the tolerances to be applied to the design

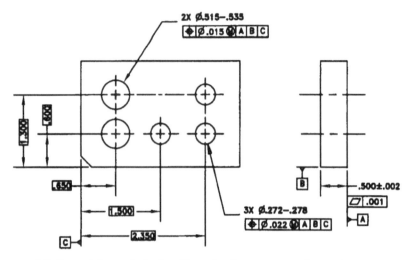

(a) Geometric controls invoking simultaneous requirements.

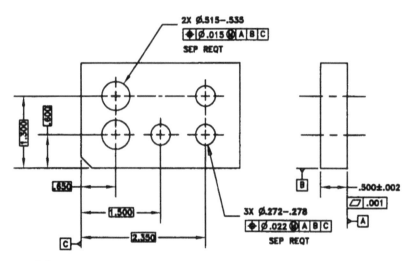

(b) Geometric controls invoking separate requirements.

FIGURE 4-9 Simultaneous versus separate requirements.

parameters can be identified and the variability of the product assessed. In a product composed of a number of components, the results of these activities will allow partitioning of tolerances and assigning them to specific elements of the assembly. These target values are the design parameters that determine functional results. They include, but are not limited to, the spatial relationships required for product operation and function. The experimental design may require modeling of components and parts at the boundary extremes (e.g., virtual and resultant conditions) as well as at nominal or target values.

In assessing target values and associated tolerances for the detailed parts, an appropriate level of resolution must be used if the functional requirements are to be met. This can be done by refining the controls and quantifying the characteristics through Y14.5M methods. The standard accomplishes two purposes. It allows precision in defining characteristics; there is no conjecture in what is actually being controlled. Additionally, it indirectly prescribes—although mostly based on surface-plate inspection concepts or parts fixtured for CMM measurement—the manner in which the magnitude of the variation might be assessed. This allows intelligent design of the product's operating characteristics and defines the variation that can be tolerated.

4.3.6 Physical Prototyping

Once the variables and attributes have been identified for experimentation, the next step is to build physical prototypes for both laboratory and field testing. These physical models implement a worst-case analysis based on virtual and other appropriate boundary conditions and introduce the necessary reality to the product. Whether computer-based techniques have progressed to a level of sophistication that makes physical models superfluous is arguable. However, physical prototypes make the interaction between the product and the process vividly apparent and can lead to the selection of target values that ensure functional, durable, and producible products.

At the risk of repeating the obvious, the use of the physical prototypes marries the designer's concept of perfection with the variation inherent in the world of manufacturing. The use of virtual prototypes (computer-based prototypes distinguished from virtual condition prototypes) does not currently allow for detailed simulation of all the possible geometric variations that may occur in production. The result of not using physical prototypes will be an increased use of engineering change orders. As pointed out, the cost in both time and money to implement avoidable changes is not acceptable.

With the advent of rapid prototyping technology, care must also be taken in interpreting results gained by using materials and techniques that differ from those used in the production version of the product. Each manufacturing process has its own signature characteristics, which may not be matched by the prototyping methods used. Risk may be increased where materials are chosen that do not match the final material choices. An example of this would be a prototype made from a polymer rather than the metal specified for the final product. It is preferable to obtain prototypes from the company that will produce the production parts, expecting that these will manifest the same variation that will be experienced in actual production. Any process or material used in the prototypes that alters these variations (both type and magnitude) and masks awareness of them introduces a business risk that will become apparent when full production is started.

Preliminary definition of each of the assembly's components is necessary to create the prototypes. The geometric controls included in the product definition place bounds on the target parameters incorporated into the prototype. The underlying rationale in constructing the model is to set its parameters at target values deemed most likely to provide useful test information. This stage of the process begins to consume significant financial and time resources that must be husbanded by an astute choice of model parameters. These values are set with the aid of geometric controls.

Accelerated life testing is an additional area where a mix of prototypes simulating worst-case parts may be used. Here, the same concern about using production materials and processes is a significant issue. Furthermore, these prototypes will not yield truly useful information unless the parts can be measured prior to and after such testing. All the above considerations should influence the experimental design plan.

4.3.7 Metrology and Product Development

As the cost of acquiring various metrological capabilities declines, it is becoming increasingly common to find that metrology assumes a much larger role in developing the product definition. Many firms are finding that this provides a strategic advantage that differentiates its capabilities and allows access to a different customer segment.

Many metrology instruments can be effectively applied to the product development process. One popular machine that augments the physical prototyping activities is the CMM. This machine tool can be used to acquire data that can be manipulated in a variety of ways to suit a multitude of purposes (Grant, 1995). Once the data set has been acquired, it can be processed by the appropriate algorithms and yield a variety of metrological characteristics for prototype and production parts. It is important to note that this data set can be acquired in a process that approaches a single setup, consequently reducing measurement uncertainty and providing more useful information.

Current CMM technology not only includes the common touch trigger probe (TTP) but also encompasses scanning probes as their price becomes more affordable. The latter devices allow high-speed data acquisition not possible with the TTP. The high data density that can be acquired in a short time period gives the inspection planner more flexibility in choosing sampling strategies to be implemented and allows him or her to capture more value from the prototype parts and processes. Past reliance on the TTP required taking fewer points due to the large amount of time needed to gather data using this traditional CMM technology. In addition, sampling speeds were usually increased with the TTP to quickly acquire the points, trading measurement uncertainty for throughput.

An example of the use of scanning technology in the development process is illustrated in Figure 4-10. The figure shows the scan of a nominal Ø.500 pin that has been centerless ground. The polar plots shown are the first [Figure 4-10(a)] and last of eight [Figure 4-10(b)] sections (i.e., levels along the part) that were scanned. As expected, it exhibits a form error that manifests itself in lobing. At each level, a total of 144 points was sampled. The figure demonstrates the change in the variation of the lobing effect and also shows that the lobes are rotating as one moves along the part.

From a functional standpoint, this pin was to fit into a nominal Ø.500 bushing. When the pin was measured using a typical two-point measurement, it yielded a value under the nominal size. If this was the only inspection measurement taken, it might be assumed that the pin and bushing should assemble. However, the scanning information shows that this is not to be the case. The presence of the lobes due to a process malfunction creates a form error of sufficient size that the function of the assembly is impaired. While a rather simplistic example, it amply illustrates how metrology may be used in the product development pro-

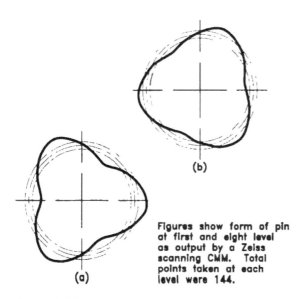

Figures show form of pin at first and eight level as output by a Zeiss scanning CMM. Total points taken at each level were 144.

(b)

(a)

FIGURE **4-10** Polar chart of results of scanned pin. (Courtesy of Carl Zeiss, IMT Corporation.)

cess to determine functional requirements that can be embedded in the product and process definition.

4.4 SUMMARY

At its most simplistic, the product design process can be described as having two major levels of concern: the design of the product's architecture and the design of the components assembled to create the product. While not the focus of this chapter, the architecture's design can be accomplished through the use of a variety of techniques. Whatever the means used, the output of this stage of product development provides specific goals for the design to achieve; these goals are assembled by translating customer requirements into specific product characteristics including quantitative measures allowing objective assessment of the design.

Once customer requirements are understood, a verbal model is created identifying the functional relationships that exist within the architecture. At this point, the subsystems that comprise the product have not taken physical form. The "how–why" relationships of the verbal model are to be used in subsequent design activities where decisions are based on functional relationships.

The substance of the chapter involves the six-step design layout methodology that creates a logical structure to guide system-level design. Once the product is sufficiently understood to start the design of the physical elements, this methodology is used to organize both the information and the decision-making processes that lead to the customer-needed, deliverable product.

The end products of this stage are a working prototype and the identity of certain problems inherent in the new design. In particular, the team will have made an assessment of the product to ensure acceptable levels of functionality, manufacturability, and automated assembly. The preliminary design layout, now incorporating the results of prototype testing, provides the starting point for the detailed component design (product/process), which the next chapter covers.

REFERENCES

Boothroyd, G., Poli, C., and Murch, L. E, Automatic Assembly, New York: Marcel Dekker, 1982.

Fowler, T. C., Value Analysis in Design, New York: Van Nostrand Reinhold, 1990.

Grady, J. O., System Requirements Planning, New York: McGraw-Hill, 1993.

Grant, M., Financial evaluations, in Coordinate Measuring Machines and Systems, J. A. Bosch, ed., pp. 149–152, New York: Marcel Dekker, 1995.

King, B., Better Design in Half the Time: Implementing QFD, 3rd ed., Methuen, MA: Goal/QPC, 1989.

Stamatis, D. H., Failure Mode and Effect Analysis: FMEA from Theory to Execution, Milwaukee, WI: ASQC Quality Press, 1995.

5

A Producible Component

5.1 INTRODUCTION

The methodology described in Chapter 4 yields a design layout that gives geometric definition to the product and incorporates knowledge gained from testing physical prototypes. In a limited sense, the end product (the system's architecture and the components that make up the assembly) is sufficiently defined to allow copies of the product to be fabricated. This cannot, or should not, be deemed a production release since no effort has been in expended in designing tooling or gaging. Both the design of the product and the production process must be completed to bring the product development cycle to a successful conclusion.

At this stage, the design layout and prototypes contain the necessary elements to complete the product development cycle; but this knowledge is not sufficient to ensure a producible design. Also, in the event that a company either wishes to retain the proprietary knowledge gained to this point or is working on a classified project, it may not be able to share the design layout with subcontractors so that they can transform the product design into the required production capability.

The design layout contains the seeds for the design of a production process where the major design criteria include producibility. The topics of this chapter deal with a second methodology (Figure 5-1) that uses the layout to develop both the tooling and gaging designs needed to create a producible product. Embedded

73

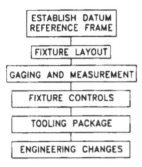

FIGURE 5-1   Six-step methodology for a producible component.

within the methodology are techniques that help resolve problems identified in developing the design layout and prototypes. The working drawings that result (e.g., product assembly, component details, gage and fixture drawings) allow production planning to specify and acquire the capabilities needed to place the product in true production.

To provide a concrete example and illustrate the various steps in the design process, the solenoid pump (see Figure 5-2), briefly described in Chapter 4, is used here. The individual elements of the assembly are discussed ahead, with numbers corresponding to the balloons in the figure. The initial version of the pump design, prior to simplification, is used to illustrate the component design process.

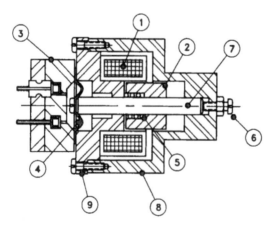

FIGURE 5-2   Pump assembly (same as Figure 4-4).

The pump is a steel-clad assembly, consisting of an electrically energized coil (1) of insulated wire that produces a magnetic field within the coil. The coil surrounds a movable iron core (2), which is pulled to a central position with respect to the coil when the coil is energized by sending current through it. The iron core is known as the armature or plunger.

The solenoid pump consists of a hydraulic loop assembly (3), a diaphragm (4), and a spring (5) that returns the armature each time the coil is de-energized and performs the pumping action. There is a screw (6) with a lock nut to adjust the flow volume of the pump. The core is mounted on a piston-like rod (7) contained in two bores. These bores are features found in the pump body (8) and the endcap (9).

5.2 STEP ONE: THE DATUM REFERENCE FRAME

As emphasized in previous chapters, the critical element in describing a component is selection of the datum reference frame. The design layout provides the relationships among components in the assembly that, in turn, allow the DRF to be identified. Hence, the layout is the starting point necessary to detail individual components.

The concurrent engineering team must determine that singular set of part features that will orient the component in three-dimensional space in its next assembly position. This set of interrelated part features is identified as the functional datum reference frame. Several combinations of features on each component of the solenoid pump could establish a DRF. A useful technique to identify the appropriate DRF for a component is to generate a list of candidate datums by looking at the surfaces where the mating components contact each other. From this list, the various combinations of these interfaces (including precedence) are collected and provide the alternative DRFs. The team must decide which of these possible choices is most critical as it relates to function and manufacturability—which one provides the component's unique location in space.

In this example, the efforts are concentrated on the pump body and the endcap. The details of these components are shown in Figures 5-3 and 5-4. Each of these has a number of features that could be used as elements in creating a DRF. The pump body has a series of three internal coaxial bores, two external diameters, eight drilled and tapped holes, and a planar surface, all of which could become elements in constructing the functional DRF. Likewise, the endcap has four parallel, planar surfaces, three external diameters, a series of eight clearance holes, and three central coaxial bores. Any of these geometric features (either singularly or in concert) could serve as an element of a DRF.

However, from both a functional and manufacturing standpoint, only one of the combinations relates the component to the assembly in the manner the

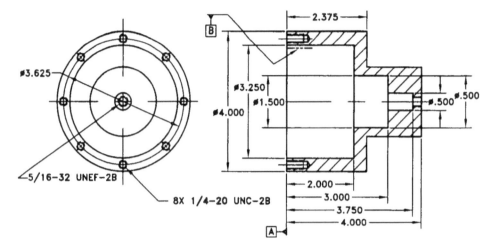

Figure 5-3 Preliminary detail for pump body (no geometric tolerance specified).

designer envisioned. This set of features, the functional DRF, will position each part with zero degrees of freedom when the part is nonaxial, or with 1 degree of freedom if it is a typical axial part.

After the list of DRF candidates for this example was evaluated, the following features that hold the two case sections together were identified as the most critical characteristics for assembly. The primary datum feature on the pump body is identified as the mating interface (the planar surface) and is designated as

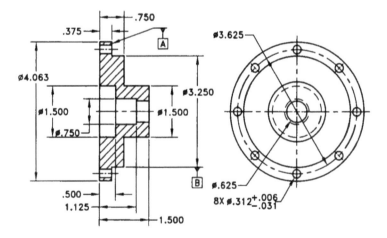

Figure 5-4 Preliminary detail for endcap (no geometric tolerance specified).

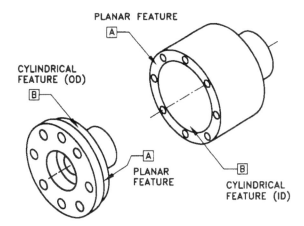

FIGURE 5-5 Pump body and endcap datums.

datum *A* (see Figure 5-5). This planar surface on the component, where it contacts a similar surface on the endcap, is the primary feature that provides kinematic orientation of the component within the assembly.

To provide satisfactory functional constraint of the part in three-dimensional space, additional datums must be identified. Based on team discussions, the secondary datum is determined to be the internal diameter on the pump body, identified as datum *B* in the figure, that locates the two case elements so that the bores in which the rod slides will be coaxial. Because of the limited penetration of the diameter of the mating cap, datum *B* is restricted to the area indicated by the chain line in Figure 5-3.

For this part, the only unconstrained motion left involves rotation about the axis established by datum feature *B*. This motion could be constrained by using the set of drilled and tapped holes that would stop rotation. However, since no asymmetrical features need to be controlled for rotational position, a tertiary datum is not specified. The hole pattern has a symmetry that eliminates the need for the part to be "clocked"; these holes could subsequently be related to the DRF by a positional tolerance.

The results of this process are documented in Figure 5-3 for the pump body. An identical procedure is applied to the endcap with similar results shown in its detail (Figure 5-4).

Once the datums are selected, they must be qualified through the use of the appropriate geometric form controls and interrelated to form the DRF. This interrelationship of the datum features through the use of controls such as orientation (e.g., perpendicularity) will avoid incomplete engineering requirements. In the example, specifications to interrelate the datums might also require locational information (a positional tolerance) if the eight-hole pattern were used as the tertiary datum, because the holes would have to be related to the other datum features. This could be done by providing basic dimensions and a positional control where the control's datum reference frame consists only of the primary datum.

In any event, when the component definition is used to develop prototype parts, all samples of the component should be built using the basic dimensions and associated geometric tolerances to determine target sizes and locations. The target values are not necessarily the nominal dimensions but should be developed using the appropriate boundary extremes that yield useful design-related information from the testing database.

The engineering team will find it impossible to complete the component design unless each part has been placed within the assembly and all the functional relationships have been defined. This argues for completing the design layout (Chapter 4) of the assembly before attempting to design or fabricate any constituent components. The release of component details to manufacturing prior to agreement on the design layout will inflict engineering changes on the process as the assembly definition is eventually completed, possibly after initiation of production. A more germane point is that manufacturing personnel may be allowed unwarranted, undesired, and possibly unrecognized latitude in completing the product definition if the team abdicates its responsibilities at this point in the cycle.

5.3 STEP TWO: THE FIXTURE LAYOUT

Using the DRF specified in the first step of component design, a fixture layout is developed that replicates the spatial location of the component in the next assembly position. In simple cases, this fixture layout will simulate the mating component at all assembly interfaces. In more difficult cases, datum features, such as tooling holes, bosses, or tabs, may need to be added to the component design, providing features of sufficient size and geometric form to allow proper part fixturing and complete kinematic control. The fixture layout then takes on additional geometric characteristics in order to provide the locating and holding surfaces necessary for workholding. A more complete discussion of fixture design is found in Chapter 12.

To illustrate how the fixture simulates the mating part, Figure 5-6(a) shows a simple assembly. Figure 5-6(b) describes the production fixture and shows how it mimics the mating component, simulating the next-higher level of the assembly. While there may be considerable work to be performed to complete the fixture design, the example illustrates the basic concepts around which a fixture can be built. Prominent among these concepts is the functional DRF with its ability to drive the fixture design. Further expansion of the principle is demon-

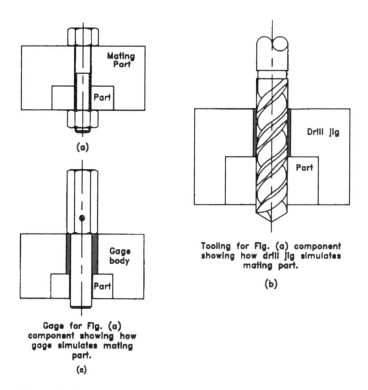

FIGURE 5-6 Example of simple design methodology.

strated in Figure 5-6(c), where the gaging fixture (functional gage) is also designed using the DRF identified in step 1.

For this pump assembly, a fixture layout is prepared for each of the components analyzed in step 1. These fixture layouts, containing both component and fixture, will replicate the next level of the assembly. As a consequence, the fixture simulates the mating part, reflecting its geometry at virtual condition, and locates each component in a singular position in three-dimensional space that replicates its actual position when assembled.

With the fixture layout in hand, the design team should add gaging elements for each of the critical and major part features that must be related to the DRF. Each of these component features is necessary to ensure part interchangeability, assembly, fit, or function. The gaging elements take their size, form, and location from corresponding features on the mating component of the part inspected by the fixture.

These fixtures are illustrated in Figures 5-7 and 5-8. In the case of the gaging fixture (Figure 5-7), the figure shows the geometric similarity between

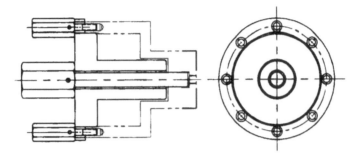

FIGURE 5-7   Functional gage for pump body.

the fixture and the mating component, particularly in the manner in which it locates the part. Using gage datums (datum feature simulators) that are identical to the locating features on the mating part, or as near to identical as manufacturing necessity allows, provides the desired location. They also become the basis for the design of the functional gages (the visible lines in Figures 5-7 and 5-8) that would be used if the economics justify the expenditure. The datum surfaces could be incorporated in a measurement fixture and used for setup purposes if CMMs or surface-plate inspection techniques are used to generate variables data.

Replacing the bushings that receive the plug gages with drill bushings alters the basic design of the gaging fixture. This creates the drill jig for the endcap illustrated in Figure 5-9. The design of the drill jig again simulates the mating component and serves to minimize possible process variation. While not a complicated example, it does show how the component design carries the seeds of the tool and gage design.

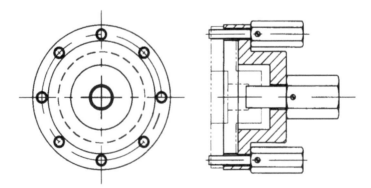

FIGURE 5-8   Functional gage for pump endcap.

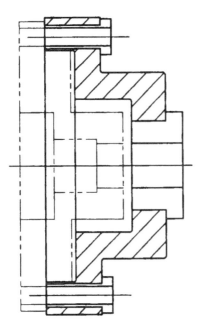

FIGURE 5-9   Drill jig for endcap derived from functional gage.

When manufacturing equipment capable of single-setup machining is available, tooling such as the drill jig may not be necessary. In this case, the exterior of the pump body could be machined using the workholding fixture shown in Figure 5-10. This would be mounted on a three-axis turning/machining center and all the critical, interrelated part features would be roughed and then finish-

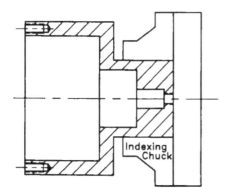

FIGURE 5-10   Workholding example for single-setup machining of pump body.

machined in this single setup. With this processing technique, all the datum features and the remaining functional features could be generated in a single setup, eliminating, or at least reducing, process errors. Since they are all generated within the single setup (a single, process DRF), the level of variation found is dependent on the machine's capability and is not influenced by multiple DRFs being chained together. Similar comments could be made for the endcap. Figure 5-11 illustrates the corresponding workholding fixture.

These functional gaging—and processing—fixtures have unique capabilities not completely duplicated by surface-plate or CMM inspection. They not only ensure the interrelationship of the datum features forming the DRF but also provide the framework within which the part features may be gaged for interchangeability. Both capabilities are simultaneously incorporated into the gaging fixture when the functional DRF is the basis for the design of the tooling. Functional fixtures provide a functional check as distinguished from a series of independent measurements (such as performed on the CMM). The latter technique does not emulate the true relationship used for verifying product function and assembly.

Part definitions that include large tolerance values relative to the feature dimension may be inspected using CMMs and the resulting information used to assess function and assembleability. As the precision level increases (i.e., tolerance values become smaller relative to the part dimension being inspected), measurement uncertainty associated with CMM discrete-point inspection methods can begin to dominate the inspection results. With extreme precision and a stable process that produces an acceptable and functional product, the CMM still proves to be a satisfactory way to monitor production. However, a point will be reached where measurement uncertainty prevents these discrete measurements from providing the desired functional check; measurement error may become the dominant portion of the resulting measurement. The measurement planner must under-

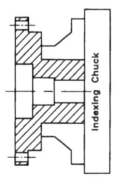

FIGURE 5-11 Workholding method for single-setup machining of endcap.

stand the manufacturing and inspection processes and select an inspection tool appropriate to the task. The selection of inspection techniques is similar to the selection of the process operations needed to fabricate the part and deserve to have the same level of importance attached to them. In high-precision situations, the discrete-point inspection measurements are useful for monitoring the process but may not ensure complete interchangeability.

The gage shown in Figure 5-7 will also determine the virtual condition of datum *B*, which includes a perpendicularity requirement and ensures that it is within the upper size limit. It could, as shown, include an assessment of the virtual condition of the largest diameter, the virtual condition of the smaller diameter, and the diameter's height. Maximum wall thickness of the case section can also be assessed.

The fixtures illustrated in Figures 5-7 through 5-11 should be concurrently designed as a set, to ensure that all four fixtures conform to the same functional design requirements. The design of the workholding fixture must complement the design of the functional gage or gaging fixture for the same component.

5.4   STEP THREE: GAGING AND MEASUREMENT

The team's next task is to use the functional DRF and the features it controls to determine whether any of the interrelated features, both critical and major, require inspection. The set of features that together form the DRF or any features referenced to the DRF may require Go, Not Go, variables, attributes, or functional gaging.

Figure 5-6 clearly demonstrates that the inspection fixture design is dictated by the choice of the DRF, as is the design of the production fixture. Structured design techniques prevent tool and gage designers from using discretion in the creation of production and inspection tooling packages that may violate design intent. The concurrent engineering team identifies and addresses the variability of the production process before hard tooling is purchased; the decisions relating to the variability are then communicated as specific design rules to be used in generating the gaging and inspection packages.

While the example in Figure 5-6 was intentionally kept simple for discussion, it is obvious that the typical industrial part is more complex. This makes the design process much more difficult. With complex parts it becomes imperative that the functional DRF is used as the starting point leading to optimum tooling and gaging designs. This crucial design focal point is to be preferred over the arbitrary choice of locating surfaces that arise when components and tooling are designed by numerous designers making individual and isolated decisions. This is an argument for an organized and logical approach to the design rather than an unorganized but possibly rapid execution of a production design that subsequently requires change.

The gaging fixtures shown in Figures 5-7 and 5-8 are the Go and functional gages based on the functional requirements identified by the team. The "function" gaged is component interchangeability based on the Taylor principle.

Variables data could be obtained from the virtual condition functional gage by adding dial indicators, air, electronic, or other metrological instruments. When variables data is required to assess process capability, the team will specify the inspection report to be used and the charting method to be implemented.

An illustration of the inspection database is shown in Figure 5-12. This table was generated to show the information required when the location of the hole pattern is measured to determine interchangeability.

The first two columns include a very unique set of x- and y-values (deviations) in inches. The hole size shown in column 3 is the largest-diameter cylinder perpendicular to the primary datum that will pass through the hole. The x and the y deviations from basic position (shown in columns 1 and 2) are the deviations of the axis of this cylinder from its basic or true position.

Column 4 is based on a calculation where the virtual condition (VC) of the hole is subtracted from the hole size in column 3. This value is the positional

Inspection Report
Product Specification: 8 x ∅ .312
(Solenoid Pump Clearance Holes)
Virtual Condition: ∅.312 - ∅.031=∅.281(VC)

Characteristic Sample Number	Column 1 X Deviation	Column 2 Y Deviation	Column 3 Size	Column 4 Size - VC	Column 5 $2\sqrt{x^2+y^2}$	Column 6 (Column 4 minus Column 5)
1	+.007	-.006	.315	.034	.018	.016
2	+009	+.008	.319	.038	.024	.014
3	-.006	+.008	.314	.033	.020	.013
4	-.006	-.007	.318	.037	.018	.019
5	-.009	+.008	.317	.036	.024	.012
6	+.008	-.006	.315	.034	.020	.014
7	+.007	+.006	.315	.034	.018	.016
8	-.004	-.006	.314	.033	.014	.019

Mean of Column 6 = .01538
Range of Column 6 = .007

FIGURE 5-12   Inspection database.

tolerance available when the hole is at its basic location in x and y. Note that the value is a function of the hole size in column 3; as the hole size increases, there is potentially more tolerance available; the opposite is true as the hole decreases in size.

Column 5 is a calculation based on the deviation in x and y for each hole. The formula, two times the square root of the sum of the squares of x and y, converts the data in columns 1 and 2 into positional tolerance diameters for each of the holes based on its deviation in x and y.

Column 6 is the difference between columns 4 and 5. The larger this value becomes, the further one is from the "edge of the cliff" of not achieving interchangeability and, therefore, assembleability.

There may be a temptation to utilize the information from the inspection report to chart positional tolerances. It must be understood that any attempt to chart geometric characteristics involves significant preliminary work to understand the underlying statistical distributions. The typical GD&T characteristic does not follow a univariate Guassian distribution, which makes investigation into the type of distribution both difficult and imperative.

5.5 STEP FOUR: FIXTURE CONTROLS

Once the inspection and measurement requirements have been determined, these relationships are embodied in geometric controls and used to complete the preliminary design layout and the working drawings. The separate fixture designs, containing functional acceptance criteria embedded in them, are used to guide the team in completing the layout. Both the layout and the fixture designs are not complete until geometric controls have been added.

Returning to the pump example, the functional features—the critical and major characteristics still contained in the design—must be incorporated into the fixture, related to the fixture DRF, and given geometric tolerances. These requirements, including any appropriate size tolerances, are the gage tolerances. Figures 5-13 and 5-14 show the results for the component details, with Figures 5-15 and 5-16 illustrating the resulting gages.

The completed design layout and the fixture drawings will be the basis for preparing the component detail drawings. At this level, the tooling package, to be discussed ahead, provides complete product definition for each component, ensuring that design intent is maintained. Additionally, the detail definition contains all the necessary information to ensure integrity of the assembly as the components are produced.

In the event that the component will be measured using surface-plate techniques, the part would be set up in a rotary table and the measurement process sheet would describe each step required to generate the database. Inspection process sheets would also describe all the necessary inspection hardware to simulate

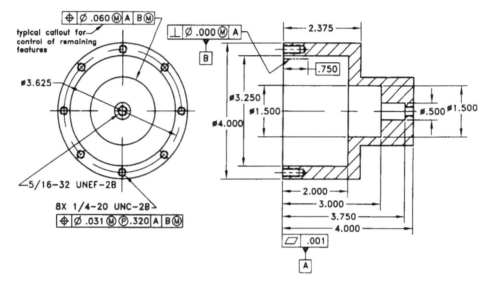

FIGURE 5-13   Detail for pump body with GD&T added.

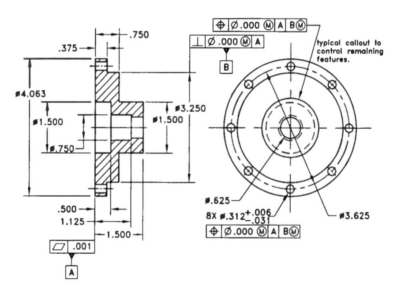

FIGURE 5-14   Detail for endcap with GD&T added.

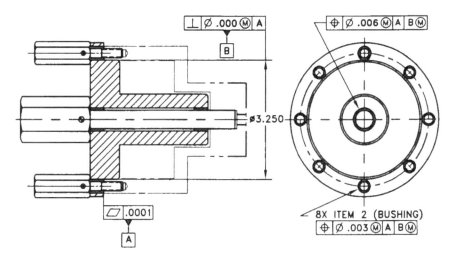

FIGURE 5-15   Pump body functional gage with selected GD&T specifications added.

the effects of the fixture. The inspection process is planned and documented to the same degree as the manufacturing process.

If measured on a CMM, the setup and sequence of operations would again be covered on a process operation sheet to ensure that the same database verifying the intended characteristics is generated. The part could be set up using standard

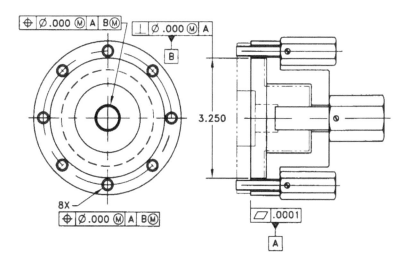

FIGURE 5-16   Endcap functional gage with selected GD&T controls added.

inspection components that can simulate the fixture from step 2 or with a fixture specifically designed—step 2 again—for measurement purposes.

Where fixture gages similar to the functional gages shown in Figures 5-15 and 5-16 are used to generate the variables database, the fixture layout includes optional sweep gages to contact those part characteristics used in determining process capability.

It should be emphasized again that checking process capability is not necessarily the same as checking for component function. The design team should engage in appropriate discussions related to the differences between these checks. In many instances, an inspection process that samples a surface by generating a point data set (e.g., a CMM) does not converge to a functional check until a large number of points are taken. With industrial emphasis on cycle-time reduction—sampling based on the minimum number of points required by the software and high sampling velocities—this divergence of inspection methods sometimes gets lost in the rush to move the part into production. Low cycle times may be obtained by increasing measurement uncertainty. This tradeoff must be consciously evaluated.

5.6 STEP FIVE: TOOLING PACKAGE

After the controls have been added to the fixture, both the component detail drawing and the fixture layout should be bundled together for manufacturing planning or requests for quotation. The team must make certain policy decisions at this point. Two explicit areas involve the scope of information to be released to suppliers and the tolerancing policy that will be applied to the inspection methods.

The first of the decisions involves how the necessary endproduct definition is provided to the supplier. If the work is to be retained by the firm and members of the appropriate departments were involved with the simultaneous engineering of the product, the design layout is available to all concerned parties. The part detail and the fixture layout of each fixture/gage (or equivalent inspection process plans for use of equipment such as a CMM), including DRFs, geometric controls, and tolerances, are sufficient to provide complete product and process definition.

Other options occur if the part is to be subcontracted. If the supplier is a member of the concurrent engineering team, possibly the result of a long-term relationship, then the package provided would be the same as mentioned above. If not part of the product development team, the contractor may not be given the design layout and he or she would not receive any of the fixture/gage layouts since these would be considered proprietary information. To ensure conformance to the product definition in such a case, the contractor would be provided with hard gages and/or appropriate inspection process plans. These gages and process plans would prevent proprietary or classified information from being released to the contractor while still providing complete definition of the product requirements.

The other decision to be made concerns the gaging or inspection policy to be employed when the gages are toleranced. The issues that must be considered in choosing a policy operate at both the economic and technical levels. At the economic level, the issue becomes the question of who accepts the risk associated with nonfunctional parts. Essentially, three policies (Galdman, 1971) may be followed:

1. The *optimistic policy*, where the buyer assumes the risk of nonfunctional components
2. The *pessimistic policy*, where the manufacturer assumes the risk of nonfunctional components
3. The *tolerant policy*, where the risk of nonconformance is shared by the buyer and the manufacturer

The technical issues (using gage design as the example) revolve around the choice of gage limits with respect to drawing limits. The technical concern is the relationship of the physical measurements embedded in the gage design with respect to the limits established by the engineering drawing. The problem requires assessing measurement uncertainty and then choosing gage dimensions that place gage limits and corresponding uncertainty zones within the range of drawing dimensions in a manner acceptable to both producer and customer—a contractual arrangement. This is illustrated graphically in Figure 5-17.

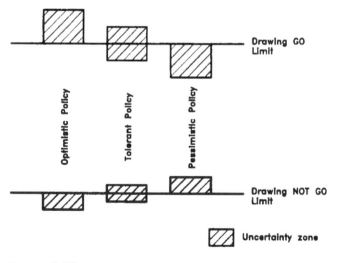

FIGURE 5-17 Relationships drawing limits, Go/Not Go gages, and uncertainty.

Just how these policy decisions could affect the development of the tooling package is seen in the situation where the components are made externally and the supplier did not participate in the simultaneous engineering activities. Hard tooling and gaging should be fabricated—again recognizing that process plans for CMM inspection may be substituted—and provided to the supplier in lieu of the tooling and gaging layouts. This ensures communicating the team's explicit definition of product conformance beyond what may be contained in the detail drawings provided for fabrication and allows no discretion in the interpretation of the controls contained in the product definition. Actual fixtures and hard gaging will ensure that the supplier understands the design requirements when bidding on the work. However, in manufacturing the tooling or gaging, gage tolerances must be established that require choosing one of the policy options. The specific gaging policy used in creating the tooling and gaging layouts must be communicated explicitly. As should be obvious, this choice has a great effect on project economics because it directly affects the choice of process and yield.

All the fixtures described or outlined above are expected, as the result of the simultaneous engineering effort, to provide the basis for production line design and quality assurance activities that ensure conformance to specification.

5.7   STEP SIX: ENGINEERING CHANGES

While one of the team's goals is to achieve production with zero engineering changes, this level of perfection is not likely to be obtained. Thus, any tentative changes should be analyzed guided by the product design (both assembly and detail level) and the fixture layouts. It is also obvious that these changes should be approved and put into effect only after the members of the original concurrent engineering team reach consensus. Any design changes will be tentatively added to these layouts and an analysis done to confirm the effects on the product design. This is done before the detail drawings are changed.

A specific example where this situation might occur is when another supplier, not part of the original process, cannot utilize the fixtures or workholding devices established by the concurrent engineering team. Under these conditions engineering changes would be required. These changes should not be approved until analyzed using the design and fixture layouts. The changes should be incorporated into these layouts to determine if the design intent, assembleability, or the producibility has been compromised. In the event that change appears necessary, the team members affected by the change must be assembled to work on the solution to the problem; the new supplier would also participate in this analysis.

5.8 SUMMARY

A critical step the concurrent engineering team must take involves refining the information contained in the design layout to give definition to the assembly's components. This definition takes the product from "functional and feasible" to producible.

The work at this stage is accomplished by a second six-step methodology. The starting point is the design layout and the working prototypes discussed in the preceding chapter. Along with these two tangible manifestations of the product, the team also identifies possible problems in the product that need to be addressed. These suspect areas are also part of the information that guides design of a producible product.

The chapter's techniques address these concerns while completing the product definition, describing in detail the components, gaging, and tooling needed to foster economical production. The result of this endeavor is the complete design layout, individual details of the product components, and appropriate designs for fixtures and gages.

The underlying idea that drives all this work is the ever-present datum reference frame. The identification of the functional frame is the starting point for team design of the fixtures and gages. Development of these designs provides the final pieces of information necessary to finish the design layout. Once this is complete, the baseline design contained in the design package—encompassing product, tools, and gages—is the benchmark from which subsequent product decisions are made.

If this process has been carried out carefully, then the inevitable engineering changes that occur will be held to a minimum. Those that must be enacted can be evaluated in a logical and controlled fashion, reducing disruption to the manufacturing operations. The use of geometric controls to define the component, the fixtures, the gages, and the production and inspection setups is an absolute necessity to achieve these goals.

REFERENCE

Gladman, C. A., Gauging principles, CIRP Annals, **XVIV**, pp. 377–384, 1971.

6

First Steps Toward Production

6.1 INTRODUCTION

The first part of the book presents foundation concepts and design methodologies to structure the product development cycle. The remainder of the text looks at critical details needed to implement the design methodology outlined in Chapter 5 and move the product into production. This chapter focuses on ideas intended to create more producible designs. Of particular interest is the contention that a component's mating part provides the design of the component's fixture and gage. In fact, in many cases the mating part is an exact image of an ideal tool or gage.

The logic contained in the following material emphasizes the design of the gage (Figure 6-1) rather than the tool. The rationale behind this is the need to identify valid functional criteria that can be used in designing both the tool and the gage; this is done most effectively by first keying on acceptance criteria embedded in the three-dimensional design of the gage. Thus, much of this chapter deals explicitly with design concerns directly related to the functional gage and inspection processes, but carries through into the design of production tooling. The impact of these design concepts on production tooling is addressed in a subsequent chapter.

In the not-too-distant past, an estimated 95% of all interchangeable parts involved patterns of clearance or tapped holes. These components were assembled with commercial fasteners such as bolts and screws. To ensure assembly,

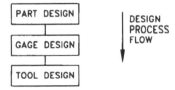

FIGURE 6-1   Producibility design cycle.

they were usually inspected with functional, three-dimensional gages designed to simulate the mating part. Even with methods such as design for assembly that emphasize reducing the number of fasteners in a product, functionally gaging parts to verify assembleability and interchangeability are still probably necessary and certainly useful in the appropriate situations.

True functional or hard gages have their limitations. The cost and time required for the design and construction of such gages can be justified only for parts produced in volume. Parts made at the beginning of production, during short runs, or as prototypes may be fixtured and inspected by open setup using surface-plate equipment. Other common methods include using optical comparators and coordinate measuring machines (CMMs). The CMM is increasingly seen as a substitute for the knowledge and skill required for open setup. The critical concern is selection of manufacturing and inspection strategies that duplicate the effects of the physical functional gage when assembleability is to be checked.

This chapter is a presentation—to product and tool designers as well as inspectors—of requirements and recommendations for ensuring that results (in particular, inspection results) are accurate, complete, and valid. It builds on the standardized methods of component design developed in the preceding chapters by which the designer can simplify drawing interpretation and increase part acceptance (yields). Subsequent chapters discuss refinements of inspection and measurement techniques and tooling design, which lead to improved accuracy and foster complete documentation of inspection results.

Applying nonfunctional inspection methods rejects many acceptable parts. Inadequate inspection techniques can give misleading results if maximum acceptance of truly functional parts is desired. The proper design and use of functional gages or equivalent methods offer the greatest opportunity to achieve the optimum economic benefit. The techniques described in this chapter highlight the sources of information used to guide both gage and tool design processes. More important, these methods verify critical information contained in the product description and infuse gage and tooling design with a consistency not found in the typical design process flow.

6.2 DESIGN

Valid inspection data begins with the product designers. Their part drawings are the basis for all decisions about setups, fixtures, and inspection methods. Incomplete or ambiguous drawing specifications force the inspector to refer to the designer for clarification or, as is usually the case, follow his or her own independent—often divergent—interpretations. For example, without specifying the DRF, there are at least 28 different ways to set up and therefore inspect the part in Figure 6-2. Each gives different results; slot 1 and hole 15, slot 1 and hole 7, slot and diameter 13, and so on.

A complete part drawing has clearly identified practical and functional datums and indicates unambiguous geometric tolerances. With this information, the inspector will know how to set up the part for inspection.

6.2.1 Datum Specification

Since inspection measurements are made to verify dimensions, both the inspection measurement and the drawing dimension should originate from the same functional datums found in the product definition. Thus datums serve a twofold purpose. As origins for the dimensions, they determine the direction and extent of the measurement. As the basis for locating the part at the next assembly level, they provide the optimum method of setup to reduce sources of error in the manufacturing and inspection. In the case where a single functional DRF is used, it eliminates tolerance stackups.

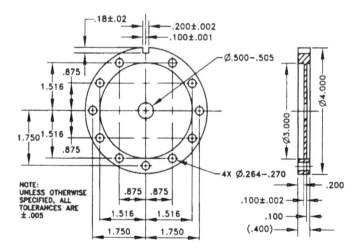

FIGURE 6-2 An ambiguous drawing using bilateral tolerancing.

6.2.2 Setups

Ideally, the setup used for inspection is derived directly from the product specification and has guided the design of the tooling used in manufacturing the part. As implied, the most effective solution to the design of process equipment is to use the functional (i.e., single) DRF contained in the product definition. Because the same xyz-axes and coordinate planes found in the Cartesian coordinate system are inherent in both manufacturing and inspection equipment, it is logical to require the designer to use this technique in specifying datums needed for processing.

6.2.3 Datum Selection

Datums are preferably chosen to identify part features that are critical for fit or function, usually surfaces that provide an interface with the mating part. However, a sufficient number of datums must be identified to place the part within a unique DRF. If a complete functional DRF was not contained in the product definition, additional, and in some sense arbitrary, datums would have to be selected. Any missing elements of the DRF will allow unconstrained motion (degrees of freedom) that can cause undesirable variation in both the manufacturing and measurement processes. It must be emphasized that this last approach (i.e., choosing nonfunctional datums) is not the preferred method of design. Ideally, the component detail will contain a complete DRF that allows no discretion in tool and gage design. The desire is to force discussion related to the DRF prior to designing hardware and implementing inspection procedures.

A cautionary note: Datums specified only for convenience of setup, or special datums that are features added for tooling and inspection, should be distinguished from functional datums by a note on the drawing. This prevents inspection measurements based on these nonfunctional datum features from being the decisive element in rejecting a part.

As mentioned, certain types of products may lack a complete DRF. If this incomplete reference frame is not sufficient to allow fabrication, then tabs, tooling holes, and similar features may be added to serve as datum features. The configuration and location of these features would be determined by the fixture and tool designer in conjunction with the concurrent engineering team. If feasible, these tooling features should be retained on the finished part to allow duplicating the setup for verification or rework.

6.2.4 Datum Qualification

Part datum features must have a sufficiently high degree of quality so that setups and measurements can be repeated and verified. The designer can obtain this by specifying close-form and orientation tolerances for these features (see Fig-

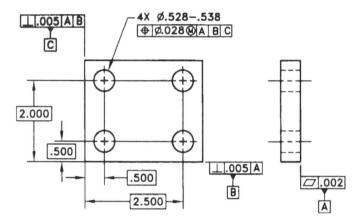

FIGURE 6-3   Form tolerances used to qualify datum surfaces.

ure 6-3). Along with the obvious need to analyze the feature's form to limit its error contribution, each of the features that comprise the DRF may need to be related to each other. This can be accomplished by the use of orientation and positional controls.

Datum targets can simulate accurate datum feature surfaces. These are indicated on the drawing (Figure 6-4) using basic dimensions. This particular application of basic dimensions (i.e., providing location for the datum targets) may have no tolerances specified or may be controlled by conventional dimensions with plus/minus tolerances. Basic dimensions with no tolerance of position attached to them are assumed controlled using gagemaker's tolerances; this invokes control based on standard shop practices for the particular type of work involved. These specified points, lines, or areas located on a datum surface contact fixturing and inspection equipment, eliminating the necessity for close tolerance control on the full datum surface, yet achieving accurate and repeatable setups. Further economy (greater process yields) can be realized if the locations of the tooling points used in manufacturing match the location of the inspection datum targets. Retaining a qualified primary datum surface (e.g., surface A in Figure 6-5) and combining it with datum targets may be desirable if based on sound functional reasons.

6.2.5   Datum Identification

A perfectly symmetrical part can still defeat the designer's intent if the inspector or setup person is unable to determine which features are those specified as datums on the drawing. A feature should be added, or the part modified, to ensure that the part is set up in only one way. In the event that a previous inspection

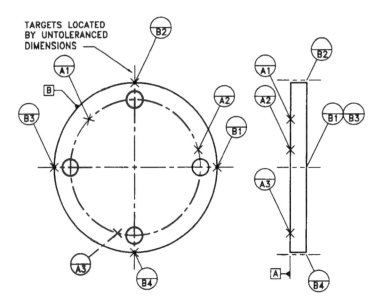

FIGURE 6-4   Datum targets.

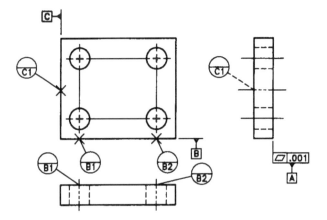

FIGURE 6-5   Combined datum targets and form tolerances.

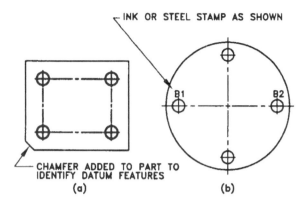

FIGURE 6-6   Datum identification using chamfering or stamping.

process must be verified, as might occur in an incoming inspection audit report, this ensures duplication of the setup. Making the part nonsymmetrical by changing the location of a hole, adding a chamfer on the corner of a nonaxial part [Figure 6-6(a)], marking datum locations on the part with ink or steel stamp [Figure 6-6(b)], or any other appropriate technique will clarify setup without being detrimental to part function.

6.3   SINGLE-DATUM REFERENCE FRAME DIMENSIONING

Dimensions are vector lines on a drawing, perpendicular to the coordinate planes comprising the DRF, and indicating by direction and distance the location of part features. Dimensions are specified in the product definition by basic dimensions and are combined with positional controls to avoid an accumulation of tolerances. In the instance where this control method is not used, the resulting tolerance stacks lead to increased process variation during manufacture and measurement uncertainty during inspection.

The recommended dimensioning technique, where feasible, is the use of a single DRF, in which all dimensions originate from the mutually perpendicular planes of the reference frame (see Figure 6-7). Using this design technique, each inspection measurement is independent of all other measurements when the features are located with positional controls: Part tolerances and errors are not cumulative. This method simplifies analysis by all concerned: from designers and process planners, quality engineers and metrologists, to suppliers. Many internal combustion engines, in particular the block and the cylinder heads, have been successfully designed using a single DRF.

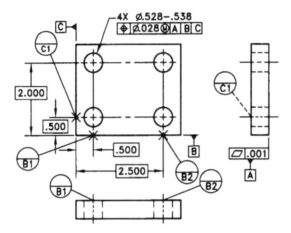

FIGURE **6-7** Datum dimensioning.

However, complications can arise when this technique is used to specify hole patterns where the pattern itself is to be treated as the functionally independent feature. Dimensioning using a single DRF may not be feasible in a straightforward manner for very large or complex parts. In these cases, multiple, chained datum reference frames or composite tolerancing may be used for proper definition. This allows flexibility in tolerance specification—and possibly manufacturing—in return for more complicated fixturing and increased difficulty in inspection procedures. In particular, multiple, interrelated DRFs would introduce unavoidable tolerance stacks; this increases process variation as errors propagate throughout the chain of related reference frames.

6.4 TOLERANCING

Tolerances are specified on a drawing for the permissible variations in size, form, orientation, and location of part features. Conventional tolerancing, widely used because of tradition, can lead to rejection of acceptable parts because they are not based on functional considerations. Even the tables of limits and fits (ASME B4 standards), which are not widely used, do not truly address the three-dimensional functionality of a product specification.

One shortcoming of these methods is the typical default decision that form errors do not contribute significantly to variation. As product precision increases (i.e., tighter tolerances), ignoring form error may no longer be acceptable in mainstream product decisions. Conventional tolerancing methods will not yield acceptable results.

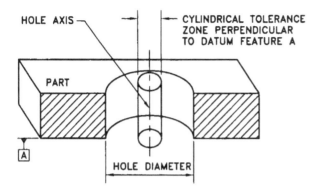

FIGURE 6-8   Positional tolerance zone.

6.4.1   Positional Tolerance Specification

Positional tolerance specifications contain the DRF from which the part features are dimensioned. The shape and size of the positional tolerance zone (Figure 6-8) are specified in the feature control frame followed by the datums in their order of precedence. This tolerance zone can be either fixed or variable in size depending on which material condition modifier is specified. The uses of MMC, LMC, and RFS conditions are covered in elementary books on GD&T. One particular application of the zone size warrants discussion because it lacks extensive usage.

6.4.2   The Zero Positional Control

Additional part acceptance can be realized when the feature location is allowed to be fully dependent on size (see Table 6-1). The true MMC size (line-to-line fit) is specified, and the positional tolerance is given as $\varnothing.000$ in. at MMC (see Figure 6-9). A part feature that is at MMC must be at true position (basic location), and a nonzero tolerance of location is allowed only when the finished size of the feature departs from MMC. Accommodation is thus made for the very best parts, those that come closest to true design specifications in all respects (zero positional tolerancing at MMC).

As a practical matter, no part will meet this standard of theoretical perfection, and a part would be rejected if it did. The latter occurs because both normal inspection policy and practice reject parts at extreme size and locational limits. However, maximum functional part acceptance is realized with the zero MMC specification because it allows manufacturing to select the working tolerance when tooling size is specified, and inspection is simplified because no separate Go gaging operation is required to inspect the MMC size.

TABLE **6-1**   Zero MMC Positional Tolerance

Finished hole size (diameter)	Tolerance allowed (diameter)
0.500 (MMC)	0.000
0.501	0.001
0.502	0.002
•	•
•	•
•	•
0.527	0.027
0.528	0.028
•	•
•	•
•	•
0.538	0.038

6.5   PHANTOM-GAGE DIMENSIONING

Phantom-gage dimensioning, a pictorial representation of zero MMC tolerancing, is a direct method of defining dimensions and tolerances on a drawing by creating a tolerancing zone for contours (profile tolerancing). It enables a product designer to pictorially describe part and inspection (gage) limit criteria on the same drawing. Following the line of reasoning presented earlier, phantom gaging also pro-

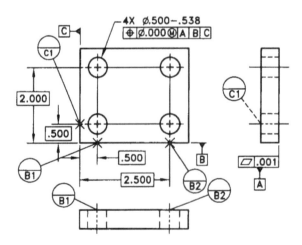

FIGURE **6-9**   Zero positional tolerancing at MMC.

vides a check on the design of the part's tooling and the mating part, as the geometric elements contained in all these artifacts are the same.

The concept is not new; an approach similar to phantom gaging was used successfully during World War II when ammunition manufacturers received gages along with design drawings. Also, the British design and drafting systems called for gage drawings to be included with design drawings. Manufacturing departments and suppliers would benefit greatly if it were common practice for design departments to provide them with design layouts.

6.5.1 Design Layout

Functional interchangeability is determined by preparing an accurate drawing of the assembly (i.e., the design layout) as discussed in Chapter 4. Since graphic design layouts traditionally show the interface relationships of mating parts, they inherently include the basic gage forms for these parts. Combined part/gage drawings illustrate these interface relationships and give direct insight into the functional requirements of the part. Figure 6-10 shows a design layout for three stacked mating parts. Each mating part is in effect also a gage.

It should be obvious that the design layout contains more pertinent information affecting tool and gage design than an assembly drawing. The assembly drawing is missing several key elements such as the identity of the DRF, assembly constraints, critical dimensions and tolerances, limits on size, and geometric controls. The following checklist may be used to extract the technical data required

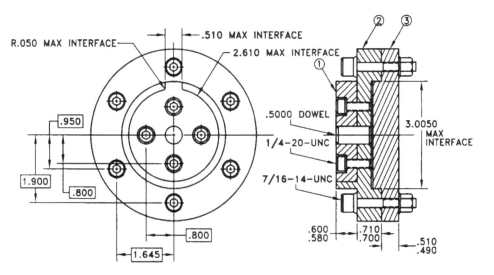

FIGURE **6-10** Design layout used in phantom-gage dimensioning.

to prepare functional tooling. This is especially helpful if no design layout has been prepared but is also an aid in verifying information that should be available from the design layout. It also helps the design team in assessing the manufacturability of the assembly.

Identification of the functional DRF.

Whether datum targets or form tolerances are used to achieve repeatability.

Must the datums be tooled with special centering fixtures (RFS) so that they cannot translate or rotate, or can they be tooled with fixed-size locators (MMC) such that both rotation and translation can occur?

If the dimensions are chained and GD&T is not used, which relationships are critical?

Are the tolerances fixed in value (RFS) or may they vary with feature size (MMC or LMC)?

What product acceptance techniques will be used for each part: optical comparator, functional gage, CMM, or surface plate?

Are the size limits and tolerances functionally derived or extracted from some ancient company standard by an anonymous author?

If the parts are flexible, can they be inspected in the free state or is a fixture required?

A third technique mentioned in Chapter 4 (i.e., no design layout before detailing) warrants discussion only to the extent that its use should be discouraged. In the event that a product is designed with no design layout, the individual components are usually assigned to different designers. Difficulty arises in communicating assembly relationships; such information is needed by independent and often isolated designers in performing their tasks. Without knowledge of the relationships, it is difficult, if not impossible, to correctly design the components in an efficient and timely manner. Conjecture is introduced into the process as the product, tooling, and gaging designers guess at the required relationships. This conjecture becomes the greatest source of inefficiency and expense.

6.5.2 Part/Gage Design Parameters

The designer establishes the following parameters on the phantom-gage design layout:

1. The method of locating parts at assembly (DRF)
2. The basic location of all features
3. The size of assembly fasteners
4. Fits and allowances, if any

The virtual condition boundaries in Figure 6-10 represent phantom-gage dimensions and form.

6.5.3 Defining Functional Gages from the Design Layout

Figures 6-11, 6-12, and 6-13 illustrate combined part and gage definitions. The phantom lines define the Go gaging (virtual condition) limits, and the solid lines represent the least critical or Not Go part limits. Since the tolerance zones are graphically represented, there is little possibility for misinterpretation. (For the sake of clarity, the phantom lines on phantom-gage drawings are not drawn to scale.)

A study of these same figures shows that only five basic rules are involved in using the phantom-gage technology:

1. The least critical (minimum material) part limit dimensions are either maximum limits for internal features or minimum limits for external features. Separate Not Go gaging operations are usually required for all these dimensions. Conventional drawing techniques are used.

2. The most critical part limit dimensions (virtual condition, usually at MMC) are enclosed in a phantom box and describe the basic Go gaging form.

3. The actual gage outline, which also represents the most critical mating part, is drawn with phantom lines. Theoretically, the relationship is

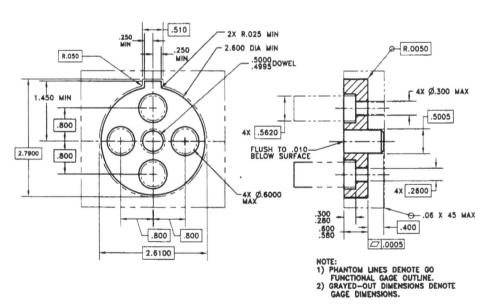

FIGURE 6-11 Part 1 detail drawing.

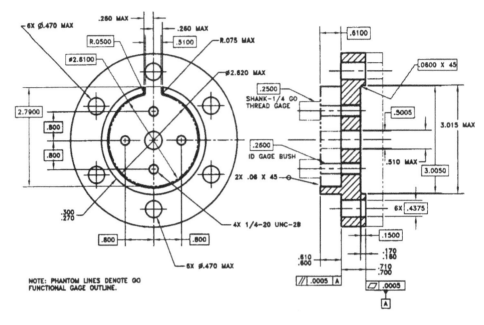

FIGURE 6-12   Part 2 detail drawing.

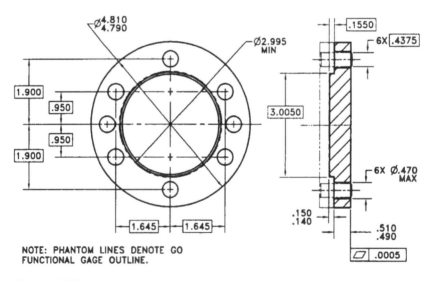

NOTE: PHANTOM LINES DENOTE GO
FUNCTIONAL GAGE OUTLINE.

FIGURE 6-13   Part 3 detail drawing.

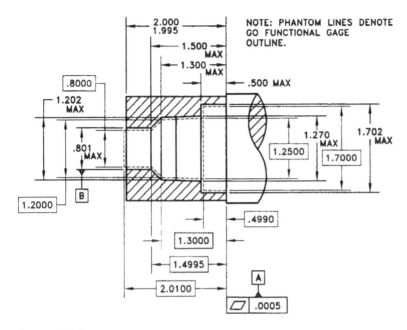

FIGURE 6-14 Detail illustration use of functional gage in defining part.

$$P_{\text{VC}} = G_F$$

where P_{VC} is the critical part size (virtual condition) and G_F is the functional (critical) Go gage size. Unilateral gage tolerances (i.e., a pessimistic gaging policy) make this Go form even more critical because they decrease internal and increase external gage feature sizes.

4. Specific gage items are individually described (see the gage bushing in Figure 6-12). However, to save space in this text, not all the gage design detail, (chamfers, radii, and so on) is indicated in Figure 6-12.

5. A minimum number of limit dimensions, form tolerances, and datum surface callouts are required. The 0.700- to 0.710-in. limit dimensions and the 0.0005-in. flatness tolerance on datum A in Figure 6-12 illustrate this type of callout. Other examples are shown in Figures 6-11, 6-13, and 6-14.

6.6 APPLICATIONS OF PHANTOM GAGING

Phantom-gage dimensioning is particularly useful for complex parts that require a great number of form, orientation, or position callouts, each requiring a detailed knowledge of drafting standards to interpret. Figure 6-14 shows a Go form that would be difficult to completely define with conventional (i.e., without GD&T) graphic techniques. The Not Go gaging requirements described in Figure 6-14 by limit dimensions are modified by the "max" callout and are inspected with separate gages or measurements. The entire Go gage requirement is checked by the single gage shown in phantom lines when the product specification invokes simultaneous requirements.

Phantom gages may also form the basis for the design of tools used to fabricate each part. For instance, if the gage pins shown in Figure 6-11 were replaced with drill bushings, the gage would become a tool. In the event that the part is to be processed on a machine capable of producing it in one setup, the knowledge gained from designing the gaging will dictate how the part is processed and what form the setup will take.

6.7   CONCLUSIONS

Phantom-gage dimensioning affects all individuals concerned with product definition, manufacturing, and acceptance. Using phantom-gage dimensioning, the product designer can graphically show tolerance limits and directly control the inspection process. One limit is the least critical part definition, and the other, the MMC limit, is the gage or mating part definition. Phantom-gage dimensioning gives manufacturing and quality engineers a graphic view of the functional requirements and highlights the importance of each part feature. This added information provides the opportunity to design a better process and associated tooling.

7

Dimensional Measurements

7.1 INTRODUCTION

Product development depends on exchanging information that at its root is quantitative and intended to represent product geometry. Earlier chapters are based on the assumption that the numbers needed to implement the design techniques were available and of sufficient quality to support the development process. To be useful, these numbers must be a valid representation of the physical features they are intended to portray. How the validity of this information might be assessed is one of the topics of this chapter.

Along with concerns relating to validity, the dimensional information must also be sufficiently accurate to support manufacture of interchangeable products. Without accuracy, the results will be suspect and may lead to undesirable or inappropriate product decisions. The development team must make measurements to obtain this dimensional knowledge. These measurements may support early stages of product development or may be useful for process development and control, final inspection, audit, or calibration purposes. For such applications to effectively contribute to the development process, the underlying theory of dimensional measurement (i.e., dimensional metrology) must be considered in implementing good measurement practice. Three major elements of good practice are covered in this chapter: general measurement theory, underlying statistical concepts, and measurement planning.

7.2 MEASUREMENT THEORY

The foundation of the measurement process is set on scientific principles, which if entirely understood and controlled would eliminate the need for statistical analysis. The process would yield a unique and valid number. However, the physical world is not quite this orderly; even the act of defining what measurement is to be taken is fraught with complexities not easily resolved. No single value emerges from the measurement process even when attempts are made to repeat the procedures under identical conditions.

7.2.1 The Measurement Model

To put the discussion on a logical footing, a model aids in defining terms and illustrating the concepts. One simple model of the measurement process is the following:

Measurement = true value + error of measurement

The true value is the theoretically correct or perfect value for the particular measurement being taken. The model implies that there is complete definition of the measurement and physically realizable procedures to obtain it. This is discussed further in Sec. 7.2.2. The true value is presumed to be deterministic in nature, having a unique value derived from its definition.

The error term can be expanded to include the following:

Error of measurement = random error + systematic error

The error terms are random variables based on probabilistic relationships governing the values they may assume. Consequently, the error cannot be characterized by a single, constant value; its nature must be captured by a statistical distribution. Furthermore, if the measurement results are to have any predictive value, the error terms must come from a process that is in a state of statistical control. A formal plan is needed that provides the necessary definitions of the measured quantity and corresponding methods and procedures to establish control. Without the measurement planning process to establish definitions and procedures, there can be no stable process and, as a consequence, no useful results.

With both deterministic and probabilistic elements providing the foundation for the model, the measured value must be described with a statistical distribution. The result of the measurement procedure is a range of values purporting to capture the true value, although some values are more likely to occur than others.

7.2.2 True Value

To clarify what is meant by "true value," we can illustrate subtleties of the underlying idea by a simplified example. An engineering drawing (Figure 7-1) provides

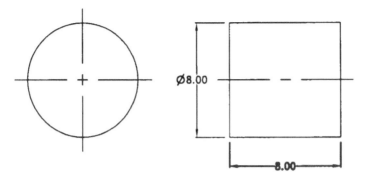

FIGURE **7-1**   Engineering drawing of simple part.

a product definition for a cylindrical pin. The physical measurement is represented by the cross-section in Figure 7-2. This illustration is one possible cross-section of many that might result from the process used to gather the data set. The figure displays the measurement results by magnifying the radial deviations from the perfect form envisioned by the engineering drawing.

The balloons in Figure 7-2 identify possible methods of describing the "size" of the cylinder. The first balloon is attached to the profile, as seen by the inspection machine, and illustrates via magnification the form deviations present in a production part. The remaining balloons describe methods that might be used to attach values to the "size" of the cylinder, none of which guarantees achieving the same numerical value or functional outcome as other methods.

Balloon 1 represents the actual profile with radial magnification.
Balloon 2 shows a two-point measurement taken over apparent high points.
Balloon 3 shows a circle whose diameter is based on an arithmetic average.
Balloon 4 illustrates the minimum circumscribed circle for the contour.

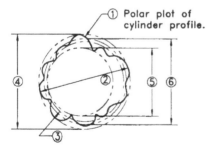

FIGURE **7-2**   Various definitions of size.

Balloon 5 is the maximum inscribed circle.

Balloon 6 is the fitted circle that might result from a mathematical algorithm such as a least-squares fitting.

From the figure it is apparent that no unique value may be attached to the "size" of this part. The mathematical complications of selecting a definition of size are further compounded by a variety of size definitions (e.g., actual size, actual local size, actual mating size) contained in the Y14.5M and Y14.5.1M standards.

The designer had a functional specification in mind when the drawing in Figure 7-1 was created. This functional specification may be better served by one of the "size" definitions here than by others. Hence, the specification must include a verification procedure to provide complete functional definition. The procedure makes explicit the choice of a size definition appropriate to the functional requirements. As an example, if a particularly loose fit is the criterion the designer used, then the two-point measurement over the high points—a local size—may be an acceptable definition. If a more precise fit is the intended design goal, the minimum circumscribed cylinder—the actual mating size—could fill the requirement, remembering that a three-dimensional fit will need a circumscribed cylinder rather than a circle for verification. Either of these might accomplish the implied functional outcome in a specific situation. However, the indiscriminate choice—or no choice—of measurement method could result in an unintended functional outcome, higher costs incurred in verifying compliance with the product definition, and ultimately unacceptable costs associated with functional failure of the product. The development team will have to decide which definition is operable within the design and economic context of the project.

An example can be used to illustrate the problem. Figure 7-3 is a polar chart of an out-of-roundness measurement for a feature that provides a high-pressure seal. The original engineering drawing addressed the issue of form control only tangentially through an explicit size tolerance ($\pm.0005$ in.) on the feature. There was no specific geometric tolerance to control roundness. Measurement planning followed the implicit instructions in the engineering specification and did not check form characteristics. The selected measurement procedure resulted in too large an uncertainty to expose the form characteristics of the manufacturing process. The unintended consequence was failure of the assembly to produce the necessary seal because the design did not establish the functional definition of the geometry. Without complete functional specification, the planning process did not include verification procedures capable of recognizing an impending functional failure (i.e., out-of-roundness).

In theoretical terms, this example illustrates the "definition problem." As can be seen, even a rather simple geometric shape such as the cylindrical pin can provide challenges to the designer of the product or measurement system. If these challenges were categorized, the two major ones would be the errors attrib-

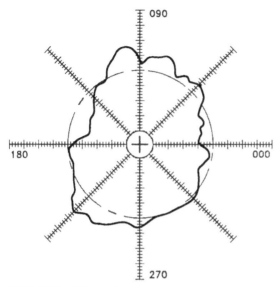

ROUNDNESS .00144 IN

FIGURE 7-3 Polar chart of out-of-roundness measurement.

utable to the theoretical description of the measurement (i.e., the definition problem) while the remaining errors are placed in the category of variation in the actual measurements.

In reality, there can never be a complete definition of the measured quantity. Too much information is required to provide complete specification of the measurement. However, like many of the processes previously discussed in the book, the definition of the measurement (i.e., the measurand) can be refined to a point where it incurs an acceptable level of cost based on design requirements and economics of the situation.

The definition of dimensional measurement to be used in this book is expansive. It includes not only the obvious ideal geometry the designer chose but also the method of measurement and the executable procedures needed to acquire and analyze the data set. Thus, the definition of the measurement is fluid, taking on different forms depending on its position in and contributions to the overall product development process. In a specific case, the planning process might lead to using a micrometer to gather two-point measurements for process control, digital sampling using a coordinate measuring machine for functional inspection, or any of a number of other techniques capable of evaluating size and other geometric characteristics. Which is appropriate is a question that must be answered by the intended use of the measurement.

7.2.3 Error and Uncertainty

In the context of physical experimentation, error means "the inevitable uncertainty that attends all measurements" (Taylor, 1982). In the measurement model, error is made up of fluctuations in measurements that may take the form of either random effects or systematic errors.

The random effects may be demonstrated by again referring to the example in Figure 7-2. Each two-point measurement taken at the same cross-section of the pin will yield a different value. Furthermore, each cross-section on the pin will also give differing measurements as the micrometer is repositioned around the perimeter of the part. The random effects can be quantified by using statistical methods to describe the dispersion of values that attends the measurement of the pin. This dispersion of values provides an estimate of the error and is what is meant by "uncertainty." The statistical property used to describe the dispersion (or variation) of the measured quantity is the standard deviation. This is the quantity ultimately used to characterize the uncertainty of a measurement, although not quite in its initial form.

The systematic component of the error (e.g., bias in measuring systems) is the "error that will make our results different from the 'true' values with reproducible discrepancies" (Bevington and Robinson, 1992). Statistical data analysis cannot identify this type of error. A statistical analysis that yields the mean and the standard deviation would be silent on the existence of systematic error. Using other measuring methods that by agreement are free from bias and comparing the resulting values with the values of interest might deal with this quandary. Another less effective technique would be to use other measuring devices and see if the values compare. The latter method may support the contention that the first device yielded unbiased results but does not prove it.

7.2.4 Precision and Accuracy

Two additional terms, precision and accuracy, are bound together much as error and uncertainty are. The term "precision" may also have a connotation similar to that of uncertainty.

Precision is the variation present in a set of measurements; it excludes the question of whether they are reasonable representations of the "true value." As mentioned, the usual choice of a statistic to describe this variation is the standard deviation. Consequently, when the standard deviation is quoted, the precision of the measurement process is being described. Continuing along this logical path, when the precision of a measurement is cited, one is also describing the uncertainty of the measurement. Actual statements of uncertainty are more involved, using a coverage factor to expand the standard deviations into a form acceptable under standardized techniques for uncertainty expressions.

The accuracy of a measurement is the closeness of its value to the true value. The model in this book implies that the true value is known, which is not possible. Much about the physical world is not yet understood, making it impossible to completely prescribe a measurement. Finding the "true value" of a measurement could rapidly evolve into a scientific exercise of suspect economics. In some instances there are conventional values that are accepted as the true value. For most manufacturing situations, a particular level of refinement (i.e., an acceptable level of uncertainty) must be explicitly chosen to implement a measurement procedure. The measurement planning process, including definition of the measurement, takes place within the development process, not in isolation. Choices are made that encompass physical principles and economics, the latter requiring detailed measurement planning.

7.2.5 Precision, Bias, Accuracy—An Illustration

Precision, bias, and accuracy are frequently used to characterize data but are applied in less than precise ways. The preceding sections have given more concise definitions of these terms, but an illustration should serve to clear up any lingering doubts as to their meanings. The classical illustration is that of target shooting with a variety of patterns serving to illustrate the meaning of each term and show the connections that must be made when choosing appropriate terminology.

The example here (Figure 7-4) uses a positional tolerance zone as the target and a single hole as the measured feature. Repeated measurements are made on this hole. Each set of measurements is made in a different setup, resulting in a specific pattern representing that set of observations. Each of these figures is assumed to have a different set of factors that affect uncertainty and provide the influences that result in the observed patterns.

The first measurement process [Figure 7-4(a)] shows a pattern with high dispersion that misses the specified tolerance zone. This pattern could result from loosely defined procedures implemented over a fairly long time span. The part is also set up on the wrong datum, introducing a bias represented by the pattern's being centered away from the target. The wide dispersion may be the result of swings in environmental factors (e.g., time-dependent temperature changes) that are not controlled sufficiently well. In this book's terminology, this pattern illustrates measurements that have high bias and low precision.

The second figure [Figure 7-4(b)] might show the same measurement process with the part correctly located on specified datum features. With no additional improvements in control of the environmental factors, the dispersion is still similar to the first measurement process, but the bias has been eliminated. This pattern shows an example of low bias and low precision.

Next, the part is measured in a well-planned process [Figure 7-4(c)]. But the inspector has located the part from the wrong datum surface as in the first example. This mistake in location has introduced bias. However, the measure-

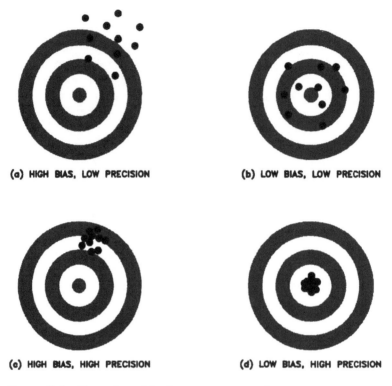

(a) HIGH BIAS, LOW PRECISION   (b) LOW BIAS, LOW PRECISION

(c) HIGH BIAS, HIGH PRECISION   (d) LOW BIAS, HIGH PRECISION

FIGURE 7-4   Bias and precision in measurement results.

ment planning process has reduced the dispersion by effectively reducing error contributors such as temperature effects. Hence, the result is a process that is highly biased but has high precision.

The last example [Figure 7-4(d)] demonstrates what happens to the measurements when all elements of the process are prescribed. The inspector now follows the measurement plan and achieves outstanding results. The part is correctly located in the specified datum reference frame, and all identified error contributors are controlled from measurement to measurement. The pattern now exhibits both low bias and high precision.

All these examples have used the words "bias" and "precision." In the world of interchangeable manufacturing, one is also concerned with being on the target. This involves accuracy. Looking back at the four patterns used to illustrate bias and precision, the last example [Figure 7-4(d)] demonstrates what might be called an accurate measurement process. It is a process in a state of statistical control that exhibits low bias and high precision. Any other combination of bias and precision does not guarantee accuracy. A process with high variability [i.e., the low precision of Figure 7-4(b)] makes it difficult to say that one is on target. A process exhibiting high bias is obviously not on target. Even if the latter process had high precision, it still does not provide accurate results. Only the combination of high precision and low bias allows a process to be described as accurate.

7.3 STATISTICAL TECHNIQUES

In the past, the gagemaker's art was sufficiently advanced relative to the manufacturing process to produce gages that utilized a small portion of the product's working tolerance. Gages could be obtained with sufficiently small errors, allowing economical control of the product. In recent years, manufacturing techniques have been developed that have process capabilities exceeding the companion measurement processes. The consequent disparity between what can be made and the ability to measure it complicates the situation. This disparity in abilities has not always been the case, nor is it the major consideration in all instances of current manufacturing practice. However, it is no longer rare to see the limitations of measurement capability having an impact in applications of high manufacturing accuracy. In such situations, it is quite conceivable that the measurement process will only recognize extreme (catastrophic) failures and only then take corrective action.

Another area where the relative capabilities of the manufacturing process vis-à-vis the measurement process causes concern is in a measurement chain where traceability is required. If traceability is based on a traditional 10% rule, each of the elements in the traceability chain has precision requirements that may rapidly exceed the capability of commercially available techniques and equipment. Consequently, it would be beneficial if methods could be found to increase the effectiveness of the measurement process without incurring tremendous cost for more accurate equipment and environments.

As it happens, there are both statistical and planning techniques that can fill the void if understood well and used appropriately. We can improve the effectiveness of the measurement process by taking the average of multiple measurements (i.e., the arithmetic mean) and using this as the estimate of the measured quantity. Such a statistical practice can increase the precision of the measurement beyond levels found in procedures dependent on a single measurement. This is accomplished by taking advantage of underlying statistical theory.

7.3.1 Statistical Concepts

The statistical techniques of use in measurement applications rest on three general ideas. The first is that the random sampling process generates values of a measurement that are repeated from sample to sample, creating a pattern that is relatively constant. One such pattern is the bell-shaped, normal distribution. Second, the sampling process provides a useful estimate—an unbiased estimator in statistical terms—of the quantity being measured. Third, the precision of the estimate is governed by the size of the sample, which is controlled by the measurement planner.

Measurement techniques rely on two different types of statistical distributions: the distribution of the individuals and the distribution of the averages. The former is the distribution of individual measurements, in theory the population of all possible measurements of the specific measured quantity that may result from the measurement procedures. The distribution of averages is the distribution of the arithmetic means that may be calculated for all possible samples of a specific size drawn from the individuals' population. The statistical name for this last description is the "sampling distribution."

The theory that underlies the sampling process has a name, the *central limit theorem*. While the intent of the book is not to teach statistics, the consequences of the central limit theorem can be illustrated graphically using Figure 7-5. In this figure, the top row of graphs shows three distinctly different populations (i.e., the individuals). Below each of the population distributions are three examples of a sampling distribution drawn from the respective population. Each sampling distribution is created by repeatedly drawing a sample of a set size from the population and calculating the sample's average. These averages are graphed and used to represent the sampling distributions.

The initial interest is focused on the location of the means of the population and its companion sampling distributions. In addition, the shape of the sampling distribution demonstrates the critical effect that the sampling process can have on estimation of the measured quantity.

Looking at the graphs, one can see that the three sampling distribution means related to a particular population example are located in the same position along the horizontal axes as the mean of the parent population. This graphical result demonstrates what the central limit theorem predicts: The mean of a sample is a good estimate of the unknown population mean. In fact, the theorem indicates that in the extreme (i.e., very large or infinite sample sizes) the mean of the population and the mean of the sampling distribution should be the same. Hence, the mean of the sample measurements can provide a good approximation (i.e., a good estimate) of one's desired measurement even without resorting to large sample sizes. It does not provide the true value, but it is reasonable to expect the calculated sample mean to lie in the neighborhood of the true value if measurement planning was done correctly. One may choose the size of the neighborhood.

Comparing the relative shapes of the sampling distributions also yields a useful fact. As seen in the figure, as the sample size is increased, the variation about the mean of the sampling distribution is reduced. From a statistical stand-

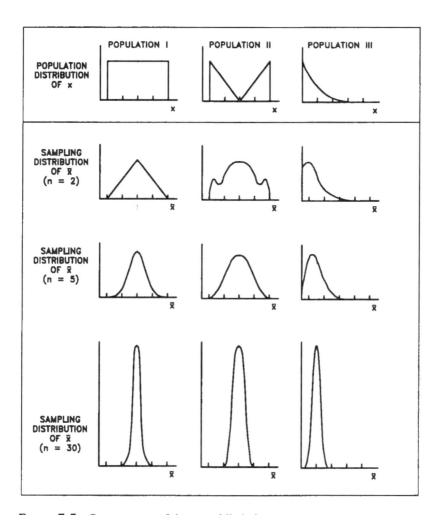

FIGURE 7-5   Consequences of the central limit theorem.

point, one gets better estimates of the mean (less variation) as the sample size increases. That this is reasonable can be demonstrated by carrying the sampling to a logical extreme. If the entire population is sampled (called a *census*), then there can be no variation in the estimate of the mean of the population because the true value has been obtained. While no census can be taken of the theoretical population of our measurement, the desired effect can be achieved by taking multiple measurements of the quantity of interest and using the average of this sample as the estimate of the true measurement. The variation about the estimate of the mean can be reduced (i.e., one can get a better estimate), but it can never be eliminated since there is no finite population of measurements from which the census may be taken. Furthermore, the cost of extremely large samples taken in an attempt to drive the variation to zero would be prohibitive.

The connection between the shape of the population and the shape of the sampling distribution should be made explicit. The central limit theorem predicts the shape of the sampling distribution to be a normal (Gaussian) distribution as the sample size becomes larger. This is also illustrated in the figure (Figure 7-5). As the sample size becomes large (the lower set of diagrams), the apparent shape of the distribution is that of a normal distribution. With the highly skewed Population III, this effect takes a little longer (i.e., larger sample size) to be achieved, but once one gets near a sample of size 30, even a sampling from the skewed population takes on the shape of the bell-shaped curve. Knowledge that the shape of the sampling distribution is normal allows one to make reasonable estimates of the confidence that one can have in the results of the measurement process.

There are limitations to the application of the central limit theorem. Knowing the shape of the sampling distribution tells nothing about the shape of the parent population from which the sample was taken. If the distribution of the individuals must be used for some design purpose such as assembly-level tolerancing, taking a sample and calculating its mean and standard deviation does not provide information about the population shape (i.e., the distribution). If the population shape is to be estimated with sufficient confidence to predict probabilities associated with assemblies composed of the individual features that the measurements represent, additional statistical investigation is needed.

7.3.2 Random Uncertainties

Returning to the measurement model here, random fluctuations are included in each of the measurements taken. The accompanying graph [Figure 7-6(a)] illustrates one source of these fluctuations by showing a statistical distribution (represented as a histogram) associated with the raw measurements. These are the uncorrected (i.e., raw) measurements taken during the actual inspection process and representing the "individuals" discussed in Sec. 7.3.1. Such fluctuations are the consequence of a host of possible uncertainty contributors discussed in the next chapter.

At this level, the information from the measurement process is a range of values that result from a sample of some set size—represented by the letter n— deemed appropriate by the inspector or the measurement planner. In theory, an infinite number of samples of this size may be drawn from the population of measurements that could be taken on this specific workpiece feature. As a consequence, the distribution shown in the figure is only an estimate of the actual population of the measurements. With care, it is a good estimator, but it is still an estimate and not the actual population distribution. So one has a little—or a

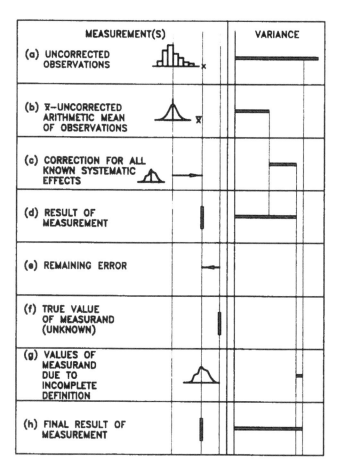

FIGURE 7-6 Illustration of measurement values, error, and uncertainty. (Follows ISO guide to the estimation of uncertainty in measurement (GUM).)

lot—of uncertainty introduced at this first stage of the process. An acceptable level of uncertainty is present if care is taken in defining the measurement; an unacceptable level of uncertainty is present if no planning is done and the measurement process is entirely uncontrolled.

Following convention, descriptive statistics are calculated for the data taken and a graphical display generated. These statistics typically include the arithmetic mean and the standard deviation. For the raw data, the uncorrected arithmetic mean [Figure 7-6(b)] is still shown in the figure with a distribution rather than a unique value, indicating the existence of uncertainty even in this value. The uncertainty arises since another sample of the measurements of the same size as the first sample will give a different value for the mean. If done correctly, the second mean will be close to the value from the first sample but will not match exactly.

Additional graphical information is added to Figure 7-6 as the right-hand graphs. They illustrate the variation in the measured values and derived quantities shown in the left-hand graphs. Such variation is represented mathematically by the standard deviation (symbolically shown as σ) or its equivalent, the variance. For the raw measurement values, the variation is the standard deviation of the sample measurements (the individuals). Once the uncorrected arithmetic mean is calculated, the statistics represent values associated with the sampling distribution and the central limit theorem. Thus, the variation for the sampling distribution becomes the population standard deviation divided by the square root of the sample size, statistically referred to as the *standard error* of the mean. The important thing to note is that the variation of the sampling distribution is less—the central limit theorem at work—than the variation in the population. This will give a tighter range of values in which the true value is likely to reside.

At this juncture, it should be apparent that the fluctuations the statistical distributions in the figure show are mathematically portrayed by the standard deviation. These values represent uncertainty in measurement. Other sources of uncertainty enter into the measurement process, but at this stage of the discussion it is important to reinforce what the term "uncertainty" represents.

7.3.3   Systematic Uncertainty

Continuing with the graphs in Figure 7-6, the mean of the raw data is adjusted [Figure 7-6(c)] for any systematic effects (i.e., bias in the measuring system). One such systematic effect could be a known temperature differential between the temperature of the workpiece and the reference temperature. While this correction might be treated as deterministic, it is likely there is uncertainty in the value of the coefficient of thermal expansion that makes a unique value unavailable. Uncertainty is just as likely in other corrections. In the figure, this underlying uncertainty associated with correction values is modeled by another, albeit smaller, statistical distribution.

It can be logically argued that the corrections can be treated as unique values (i.e., deterministic) rather than following statistical distributions. Such a decision can be based on the systematic component's being small by comparison to total process uncertainty. This is another manifestation of the refinement concept introduced early in the text. The degree of refinement necessary to achieve the project's uncertainty budget is one of the many decisions that must be made in bringing about economical production. The expense of determining the distribution underlying a systematic uncertainty contributor may be unwarranted if the correction makes a very small contribution to total uncertainty. Using a deterministic correction, the results are displayed in Figure 7-6(d).

7.3.4 Uncertainty in Definition

Preceding sections allude to the importance of the measurement definition. As pointed out, the definition given here encompasses more than normally considered when defining a measurement. It includes not only drawing representations of the ideal geometry associated with the measured feature but also methods and procedures to be followed in measuring the actual geometry and attaching a value to the measurement.

Reducing definitional uncertainty requires as complete a specification of the measurement, method of measurement, and procedures as is possible within the context of scientific knowledge and available economic resources. For example, one may choose a definition that does not incorporate all of one's knowledge, deferring to economic considerations to obtain a measurement that is "good enough." Under any circumstances, this argues for a formal measurement plan rather than defaulting to the individual inspector's judgment.

The definitional uncertainty is shown in the accompanying graph [Figure 7-6(g)] by the true value and a spread of values around it. If one cannot completely define what measurement is to be taken, then one cannot specify what the true value of this measurement should be. Uncertainty occurs because one cannot specify all the necessary details to achieve the true value. This may appear to be a purely academic point, but one very important conclusion should be drawn from this. There must be a formal measurement planning stage in every product development project to avoid introducing unexpected uncertainty into metrics used to make engineering and management decisions. If one does not know what is actually being measured and how the values are acquired, then the resulting information is useless. It is of no economic value.

7.3.5 Averages and Individuals—An Example

To show how the sampling process works to one's advantage, an example has been constructed using statistics taken from a real data set. In this case, a series of measurements was made on a composite sheet that had a thickness specification of 0.0430 to 0.0470 in. The sample measurements gave a mean thickness of 0.0449 in. and a standard deviation of 0.0014 in. These numbers are referred to as "sample statistics" to set them apart from the "population parameters" describing the theoretical population.

For the simulated example, a theoretical population that has the same mean and standard deviation found by the actual sampling process has been constructed. This theoretical population is modeled as a normal distribution, which may not be the actual case but serves to illustrate the general idea. This theoretical population will now be used to compare what happens when an individual measurement is used to represent a "true value" and, alternatively, what results if a sample of some size other than 1 is used to the same end.

The simulation uses a commercial statistics package to randomly generate 100 sample points. These are drawn from a theoretical population with the mean and standard deviation mentioned. The resulting sample of 100 points is shown in Table 7-1. Two graphical representations of this data set are shown in Figure 7-7. The right diagram is a box-and-whisker plot and the left diagram is a histogram. The general shape of the histogram should reflect the normal distribution expected from the sampling process, and the sampling distribution's bell shape should be further enhanced since the assumed parent population is also normal.

The figure shows added information in the form of descriptive statistics calculated for the sample. The sample of size $n = 100$ gives a sample mean of 0.0450 in. and the standard error of the mean (the sample standard deviation divided by the square root of the sample size) of 0.0001 in., a very good estimate of the assumed population values.

If one uses the sample points in the table and looks at these values as

TABLE 7-1   Sample Data Set for $n = 100$

Sample	Thickness	Sample	Thickness	Sample	Thickness	Sample	Thickness
1	0.046292	26	0.046455	51	0.044547	76	0.045082
2	0.046062	27	0.04652	52	0.044762	77	0.046657
3	0.046067	28	0.043585	53	0.045516	78	0.044588
4	0.044504	29	0.043611	54	0.045105	79	0.045054
5	0.046633	30	0.046825	55	0.045577	80	0.043194
6	0.046460	31	0.043823	56	0.044179	81	0.046135
7	0.043619	32	0.045327	57	0.045472	82	0.045219
8	0.044431	33	0.045252	58	0.042553	83	0.044584
9	0.042625	34	0.044787	59	0.044596	84	0.047843
10	0.043773	35	0.044548	60	0.045728	85	0.045038
11	0.045980	36	0.041770	61	0.045949	86	0.044053
12	0.045277	37	0.047336	62	0.044467	87	0.046354
13	0.045569	38	0.045978	63	0.045089	88	0.046584
14	0.043898	39	0.045020	64	0.046888	89	0.047357
15	0.043275	40	0.047223	65	0.046738	90	0.047250
16	0.044570	41	0.044123	66	0.042545	91	0.042226
17	0.044681	42	0.045378	67	0.044482	92	0.041004
18	0.044939	43	0.047079	68	0.045068	93	0.045434
19	0.043421	44	0.044608	69	0.044453	94	0.041805
20	0.045920	45	0.044913	70	0.043977	95	0.046173
21	0.045358	46	0.047167	71	0.045192	96	0.047763
22	0.045394	47	0.043874	72	0.043530	97	0.045182
23	0.045073	48	0.044463	73	0.044715	98	0.045457
24	0.044076	49	0.045220	74	0.049442	99	0.043889
25	0.045394	50	0.044322	75	0.043238	100	0.044502

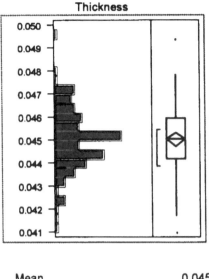

Mean	0.0450
Std Dev	0.0014
Std Error Mean	0.0001
N	100

FIGURE 7-7 Graphical display of sample data set for $n = 100$.

individuals, a different story emerges. While the histogram shows that it is much more likely to take an individual measurement and find this value near the population mean of 0.0449 in., other possibilities may occur.

The two points on the box-and-whisker plot that stand out from the rest of the diagram are called *outliers*. These points are located at measurements of 0.0410 in. and 0.0494 in., well outside the specification limits. Now consider the case where a single measurement (an individual) is used to assess whether the process is operating at its target value. Further assume that the target is the mean of 0.0449 in. A sample point at either one of the extreme values would lead one to believe that the process is not centered properly. However, this little experiment has set the process on its target. So the process is centered where it should be although process variation is greater than desired. If centering adjustments are made based on one of the extreme values, the process mean, which is at target, will be moved. This will throw more production outside the specification limits than is already the case.

It is not necessary to take large samples to achieve the desired sampling effects. Again using the data points in Table 7-1, a random sample of size $n = 10$ is shown in Table 7-2. These 10 samples were randomly selected from the

TABLE 7-2   Sample Data
Set for $n = 10$

Sample	Thickness
1	0.046738
2	0.045082
3	0.044913
4	0.045073
5	0.044053
6	0.044939
7	0.045577
8	0.045277
9	0.045358
10	0.047763

Data taken from Table 7-1.

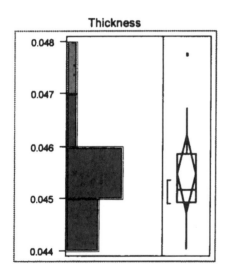

Mean	0.04548
Std Dev	0.00105
Std Error Mean	0.00033
N	10

FIGURE 7-8   Graphical display of sample data set for $n = 10$.

original list of 100 measurements. The sample points are graphically displayed in Figure 7-8 along with their sample statistics. As the central limit theorem predicts, the sample mean is still a good estimator of the population mean (0.0455 versus 0.0449). The standard error of the mean—the sampling distribution's variation—is shown to be higher than the equivalent statistic from the larger sample (0.0003 versus 0.0001). However, remember that taking a larger sample can control this.

The key point is that the sampling process—with $n > 1$—gives a better estimate of the population mean than the individuals and allows one to describe confidence in the estimate. The latter concern cannot be addressed if a single value is used as the estimator. Without a range of values to measure the variation, one cannot attach any confidence level to an estimate. A sample size of 1 is not of much value.

7.4 MEASUREMENT PLANNING

In product development the sequence of events starts by taking customer requirements and reducing them to product architecture. This stage of the process is examined earlier in the book. Product characteristics derived from the architecture—the interest here is primarily the geometry needed to implement customer requirements—must be translated into specifications. These specifications are then used for both product and process development. Measurements may be used for a wide variety of purposes in the development process: to determine specifications, monitor manufacturing process, and to verify whether design intent has been faithfully executed.

With engineering specifications in hand, a series of standards can be used to illustrate the logic that underlies the measurement planning process. In discussing this, the broad view common throughout the book is again taken. Measurement planning begins with specification of design intent on the engineering drawing and continues through the more obvious steps associated with selection of measurement methods and procedures.

The following material presents a formal measurement planning process. The need for planning can be highlighted by examples showing the relative effects of different elements of the measurement process. Based on the authors' experience, only a small portion of the uncertainty of the measurement process is contributed by the measuring instruments. Other sources estimate that 10% to 20% of the uncertainty on the shop floor is attributed to the instrument calibration (Bennich, 1997). Larger effects are introduced by temperature (e.g., possibly three to four times larger) than by the calibration of the instruments. All the critical process elements (e.g., environment, setup, definition of the characteristic, etc.) must be managed to achieve a stable and controlled process. This state of

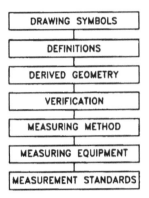

FIGURE 7-9   Chain of standards for characterizing product. (From ISO/TR 14638.)

statistical control, necessary to produce valid results, is obtainable only through a formal planning process.

The discussion can be structured using a chain of standards. This is the generic title applied to the group of standards that underlie the measurement process. The chain links are codified in an ISO technical report, ISO/TR 14638, and illustrated in Figure 7-9. The report deals with a plan that describes a geometrical product specification (GPS), allowing one to geometrically characterize the product. In the following discussion, the various chain links are referred to in more general fashion as steps. The intent is to convey more of the substance of the measurement planning process and less of the details of the GPS masterplan and underlying standards.

Reaching back to earlier ideas, the GPS standards support a design methodology and provide a communication aid for the dimensional planner to use. Without the rigor imposed by a formal process that uses the standards chain, one has an uncontrolled measurement process. This produces unusable values and moves design decisions downstream where design oversight is lacking.

A rather dramatic reason to encourage measurement planning is also cited by Bennich. In an examination of a particular country's firms, it was discovered that the typical uncertainty of measurement was greater than the drawing tolerance. This is essentially the same experience the authors have had with domestic firms and adds to the importance of the following discussion.

7.4.1   Functional Representation/Design Intent

Using the chain-of-standards analogy, the initial links are charged with representing the designer's functional intent. This is accomplished by standardized methods that indicate functional requirements on the engineering drawing. Selecting the Y14.5 standard as an example, one section in the standard deals with symbology and syntax (step 1) used to specify geometric characteristics and dimensional requirements. Other sections of the same standard—and ASME Y14.5.1—then define the theoretical intent (step 2) of these drawing indications, providing definitions for specific types of part feature geometry and associated tolerance zones.

The chain-of-standards concept is a formal recognition of the definition problem highlighted earlier. For a measurement to be used in prediction and decision making, an adequate definition must be assembled that prescribes all necessary elements needed to achieve the desired level of accuracy. Each of these chain links contributes to building a definition supported by appropriate methods and procedures that underlie good measurement practice.

7.4.2 Derived Geometry

The next steps in the chain describe specific methods and procedures used to take the drawing's representation of design intent and apply it to the physical components. Step 3 recognizes that measurements may be taken in such a form that additional processing of the raw measurements is needed. This added processing derives the quantity that will be compared to the drawing specification. An obvious example in this area would be using the CMM to determine size and a host of other geometric characteristics. The CMM addresses design characteristics—designer's intent—by a point data set that must be mathematically manipulated to give results that can be compared to the drawing specification. The geometry of the real part must be reconstructed or derived from the data set. The derived geometry can be a boundary model of the part fitted to the data set or a feature such as an axis—resolved geometry in Y14.5.1—used as a proxy for the desired perfect geometry.

The derived geometry is defined in such terms that appropriate methods, procedures, and algorithms may be identified for the measurement task. In particular, a specific geometry-fitting criterion is to be selected based on functional requirements. Fitting algorithms such as least squares, maximum inscribed circle, minimum circumscribed circle, and mini–max can all yield different results depending on the nature of the measured part. This is of particular concern when dealing with parts that have significant form errors. Each fitting criterion may cause widely differing results and functional failure due to the manner in which the algorithm creates the derived geometry (Phillips, 1995). Precautions must be taken when selecting the fitting criterion, ensuring that the substitute geometry verifies the intended part function.

7.4.3 Conformance

Given a definition of the characteristic and its derived geometry, procedures are established to compare either one measured value to another or a measured value with its tolerance. Because there is uncertainty inherent in all measurement processes, this step (step 4) is necessary to explicitly define the rules used in evaluating conformance. Guidance to be used in creating these rules can be found in ISO 14253-1, which provides decision rules for proving both conformance and nonconformance.

An important distinction must be made when deciding whether a workpiece conforms to specification. The tolerance zone created by the product specification and the conformance zone that results when uncertainty is considered are not identical. Furthermore, the size of the zones where conformance can be proved and where nonconformance occurs are subject to measurement uncertainty, a variable quantity.

Using a two-sided specification zone as an example, the effects of measurement uncertainty can be illustrated by Figure 7-10. The introduction of uncertainty into the measurement process reduces the specification region, creating a conformance zone that lies within the specification limits. If the measurement is modeled as a measured value with uncertainty equally disposed about this value, it is apparent that the nonconformance region is also affected by uncertainty.

Taking the measurement and its associated uncertainty and moving it along a line representing the specification shows the results. In Figure 7-10(a), a measurement is acceptable as long as it never gets closer to the upper or lower specification limits than the boundaries created by the uncertainty analysis.

If the measured value becomes coincident with either of the specification limits [Figure 7-10(b)], the specification zone is reduced by one half the width

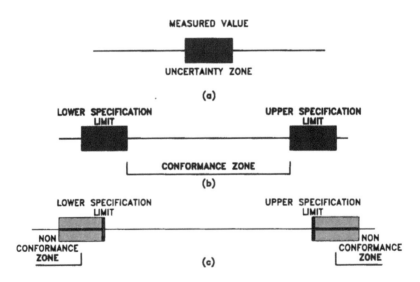

FIGURE 7-10   Measured values, uncertainty, and specification limits.

of the uncertainty zone. This width is the "expanded uncertainty" defined in ISO 14253-1 and creates the conformance zone.

When the measurement is outside the specification zone, another problem is encountered. As shown in Figure 7-10(c), the measurement is clearly out of specification when it is past the upper specification limit by a distance equal to the expanded uncertainty. However, if the measurement is near the limit and within a distance equal to the expanded uncertainty [Figure 7-10(b)], nonconformance cannot be proven. One can prove nonconformance only when the measurement is outside the specification limits by more than the expanded uncertainty.

As a consequence, there is a region where one cannot prove or disprove conformance of a measurement. This quandary arises when the uncertainty straddles the specification limit [Figure 7-10(b)]. To resolve the problem, the development team must reconcile two views of the product world. The first view is the designer's, with sharply defined specification limits that describe design intent. The competing view is that of the metrologist, where conformance limits are established by the variables of the measurement process. In the latter world, there are no fixed limits and the variability is a function of the methods and procedures used in the measurement process.

The general rule stated in ISO 14253-1 is "the uncertainty of measurement always counts against the party who is providing the proof of conformance or nonconformance and therefore making the measurement." The party proving conformance or nonconformance uses his or her actual uncertainty of measurement in making a decision.

7.4.4 Methods and Procedures

One must next choose an appropriate measurement method and select the necessary metrological equipment to implement it. Step 5 provides the theoretical structure within which the method should be selected. Step 6 provides the characteristics of the measuring instruments that will be used to implement the chosen method. The actual chain link encompasses only the metrological characteristics that influence the uncertainty of the measurement. There are also general design characteristics of the measuring equipment that are considered in assembling the instruments and selecting procedures to be followed. More is said about this aspect of planning in the next chapter.

7.4.5 Link to System of Units

Since interchangeable manufacturing and commerce require accurate measurements, the last step (step 7) defines the characteristics of a calibration process needed for the selected measuring equipment. This may include characteristics of the physical standards used in calibration procedures.

The calibration process provides traceability of all measurements made on the manufactured part to the national and international standards of length. Embedded in ISO documents is the requirement that this traceability be more than just a paper trail of documents that lead from the shop floor to a national laboratory. Statements of uncertainty are now required in each link of the traceability chain, with supporting evidence to show how these estimates were derived.

Since measurements on a manufactured product can encompass many types of geometric features, a variety of paths may be used to provide proof of traceability. These may range from the use of calibrated end standards, form standards, or mastered prototype parts. Establishing standards by the use of prototypes is exemplified by involute gearing. In some measurement applications in gear manufacture, a master gear is used as the standard. In the event that a calibrated master gear is used, a standardized procedure is necessary to calibrate the standard and demonstrate the unbroken traceability chain along with appropriate expressions of uncertainty.

7.5 SUMMARY

The continuing technological change in manufacturing precision has placed tremendous pressures on measurement and inspection processes. It was once possible to create gages of sufficient accuracy relative to the process they were intended to monitor so that no formal planning process was needed. However, with today's manufacturing processes now capable of precision that matches measurement capability, more thought is needed to provide useful information for decision making.

Three tools are presented in this chapter that can provide the foundation from which the new challenges may be met. Of a general nature are the introduction to the measurement model and the underlying statistical basis. An understanding of the place that each assumes in verifying whether a product conforms to specification is necessary before delving into the details of planning an actual measurement process. Of greater impact is the measurement planning process itself.

It has been a fairly typical occurrence to find process plans that included a single, ill-defined operation calling for "inspection." By default, the actual planning of the measurement process was left to discretion of any inspector that might be assigned the task. In past decades relatively coarse product tolerances allowed this haphazard method to yield economically viable accept–reject decisions. This is no longer the case.

A formal measurement planning process provides the control and consistency needed in the global environment, where manufacturers are being held to higher standards. The philosophy incorporated in the ISO 9000 series of standards demands a formal recognition of the elements of the product development process, including the measurement planning stages. The presentation in this chapter emphasizes the basics that must be understood to meet the more stringent requirements now being forced on manufacturers by technological advances and market requirements.

REFERENCES

Bennich, P., Seminar on Uncertainty, presented May 17, 1997, Gaithersburg, MD, under the auspices of the ASME.

Bevington, P. R. and Robinson, D. K., Data Reduction and Error Analysis for the Physical Sciences, New York: McGraw-Hill, 1992.

ISO, Guide to the Estimation of Uncertainty in Measurement (GUM), 1st ed., 1993.

ISO/14253-1: 1998, Geometrical Product Specifications (GPS)—Inspection by Measurement of Workpieces and Measuring Equipment—Part 1: Decision Rules for Proving Conformance or Non-Conformance with Specification.

ISO/TS 14253-2: 1999, Geometrical Product Specifications (GPS)—Inspection by Measurement of Workpieces and Measuring Equipment—Part 2: Guide to the Estimation of Uncertainty in GPS Measurement, in Calibration of Measuring Equipment and in Product Verification.

ISO/TR 14638, Geometrical Product Specifications (GPS)—Masterplan.

Meadows, J. D., Measurement of Geometric Tolerances in Manufacturing, New York: Marcel Dekker, 1998.

Phillips, S. D., Performance evaluations, in Coordinate Measuring Machines and Systems, J. A. Bosch, ed., pp. 149–152, New York: Marcel Dekker, 1995.

Taylor, J. R., An Introduction to Error Analysis, Mill Valley, CA: University Science Books, 1982.

8

Inspection and Verification

8.1   INTRODUCTION

Even though he or she is given a part drawing with complete and unambiguous specifications, the metrologist must contend with problems of uncertainty introduced by the equipment and the inspection procedures. The metrologist or inspector must properly align the part during setup; make a complete set of measurements; verify and allow for inaccuracies in equipment, environment, and measuring procedures; and ensure the repeatability of inspection results by recording setup and measurements fully on the inspection report. As a rule-of-thumb, all this must be accomplished so that measurement uncertainty generally will not consume more than 10% to 20% of the part tolerance.

The phantom-gage dimensioning concepts presented in Chapter 6 free inspection and quality control personnel from decisions based on conjecture that could adversely affect final product acceptance. A part defined using the phantom-gage technique can be inspected in at least one of five different ways:

1. With three-dimensional functional gages
2. With optical comparators using chart gages
3. With paper layout gaging techniques
4. With surface-plate techniques
5. Through coordinate metrology (e.g., coordinate measuring machines)

Each inspection technique has its own unique characteristics and areas of application. Hard gaging allows accurate confirmation of assembly and interchangeability. The other techniques provide more limited understanding of a part's interchangeability but prove useful in monitoring process or assessing individual characteristics. Design and application of functional receiver gages are discussed in Chapters 9 and 10. The remaining techniques are covered in Chapter 11. But before going into these details, some broad inspection topics need to be discussed. Knowledge gained from these topics should be part of developing any measurement plan regardless of the actual inspection techniques used.

8.2 PROCESS PLANNING

In the past it was not unusual to see inspection given the same weight in the planning process as material handling. Examples abound of process plans with a single operation containing the solitary word "inspect." It was a different time with different employee skills and different contractual requirements. If inspection included gaging, one could assume with confidence that the gage was sufficiently accurate with respect to product tolerances and that uncertainty in measurement was negligible—negligible in the sense that gage accuracy did not have a major influence on the overall acceptance rate.

The manufacturing world has changed significantly. Rapid increases in machine tool capability have approached precision levels exceeding the ability to measure. It has been estimated that tolerances are decreasing in size by a factor of 3 every 10 years (Lorincz, 1993). With this increase in technological capability has come the obvious request for tighter product tolerances. The result is that measurement uncertainty now becomes a major participant in accept–reject decisions and resulting contractual disputes.

To address the underlying issues, the American Society of Mechanical Engineers (ASME) created a voluntary standard, B89.7.2, Dimensional Measurement Planning. The standard seeks to eliminate the vague "inspect" operation found on many older process plans. When invoked by contract, measurement planning requires the manufacturer to prescribe in detail what is being measured and how it will be done. The latter item requires formal specification of methods, procedures, algorithms, and decision rules to meet a specific business need. In other words, you have to put as much effort into planning the measurement process as you do the manufacturing process. Furthermore, you have to establish the ground rules to verify conformance prior to encountering contract disputes.

8.2.1 Process Variation

Measurement procedures must be more accurate than the manufacturing process in which the measurements are embedded. This is due to the uncertainty contributions of the instrument and interactions between the measured entity and the metrological equipment. Such a dictate is increasingly more difficult to achieve as manufacturing technology continues to improve. This book emphasizes that two worlds are associated with product development: the virtual world of the designer and the physical world of manufacturing and metrology. During the development process, these two worlds converge. If this does not occur, designers are inclined to request product tolerances that are too tight to be measured, from either a physical or economic standpoint. Using the 10 to 1 ratio, which is sometimes referred to as the "gagemaker's" rule, a tolerance of 0.0005 in. would require a gage tolerance of 0.00005 in., which could be well outside the measurement capability of the average shop. Blind adherence to this 10% rule is likely to inflate inspection costs, in many instances well beyond what is necessary.

At the other extreme, dismissing the intent of the 10% rule leads to measurement applications where the results border on the ridiculous, containing such large uncertainty that the numbers are meaningless. One specific example involved using a ball-micrometer to measure the thickness of an iron oxide film placed on a glass tube used in electronic products. The drawing specification called for a dimensional magnitude on the order of 0.001 in., which was not likely to be verified by the measuring technique chosen.

During the design process, the total variation that can be tolerated in the product is apportioned—sometimes referred to as "error budgeting"— to individual stages of development. This includes a target uncertainty estimated for the measuring process. A simplified example illustrates the impact of error budgeting on measurement planning.

The example starts by viewing the manufacturing process as consisting of manufacturing and measurement. The total process variation observed is the combined effect of manufacturing and measurement variation:

$$\text{total product variation} = \text{manufacturing process variation}$$
$$+ \text{ measurement process variation}$$

In general, these effects depend on the underlying statistical distributions, and the results are not likely to involve simple arithmetic operations. Deriving valid statistical representations for an application requires knowledge of both statistical theory and the real process distributions. Statistical techniques used where there is significant departure from a normal distribution can lead to complex derivations and are not covered here. An extensive presentation of the theory can be found in Haugen (1968), particularly covering the algebra of normal distributions.

The simple example here follows a presentation by Nilsson (1995). It assumes that variation for the manufacturing and measurement processes is modeled by two normal distributions. This situation yields simple mathematical rules to combine variation. The normality assumption leads to an arithmetically manageable method to statistically determine the effects of the manufacturing and measurement processes, giving estimates of their respective contributions to total process variation. Without delving into the algebra of normal distributions, one can say that the total process variance is the sum of the manufacturing and measurement variances:

$$\sigma^2_{total} = \sigma^2_{manufacturing} + \sigma^2_{measurement}$$

Since the standard deviation is the square root of the variance, the standard deviation (σ) of the total process is the square root of the sum of the individual standard deviations squared:

$$\sigma_{total} = \sqrt{\sigma^2_{manufacturing} + \sigma^2_{measurement}}$$

The preceding chapter shows that the standard deviation forms the basis to calculate uncertainty values. Returning to the model, the statistical addition of the standard deviations from normal distributions is of the same form as the Pythagorean theorem. Consequently, one can use a right triangle to visually describe the logic of the uncertainty calculations. Figure 8-1 illustrates the relationship between total product variation (the hypotenuse of the triangle) and the statistical sum of the variation associated with the elements (i.e., manufacturing and measurement) of the simplified product development model. In the figure the legs of the process triangle represent the variation in the manufacturing and measurement processes.

Both the manufacturing and measurement processes have a capability that can be described using their natural limits. These limits express the natural variability observed for either manufacturing or measurement and are generated by a random process operating within a state of statistical control. The effective variation of these values is captured by the standard deviation and illustrated in the figure by the lengths of the sides of the triangle.

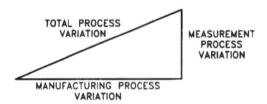

FIGURE 8-1 Relationship among manufacturing, measurement, and total process variation.

The triangle in the figure is drawn with the horizontal leg twice the length of the vertical leg. The intent is to graphically demonstrate the effects of the measurement process on the resulting total process variation when measurement variation is equal to one half the variation due to manufacturing. Using the relationship displayed above to calculate the total observed variation yields

$$\sigma_{measurement} = .5\sigma_{manufacturing}$$

$$\sigma_{total}^2 = \sigma_{manufacturing}^2 + \sigma_{measurement}^2$$

$$\sigma_{total}^2 = \sigma_{manufacturing}^2 + (.5\sigma_{manufacturing})^2$$

$$\sigma_{total}^2 = 1.25\sigma_{manufacturing}^2$$

$$\frac{\sigma_{total}}{\sigma_{manufacturing}} = 1.118$$

The combined effect of variation from both manufacturing and measurement results in only a 12% increase in observed total process variation when the manufacturing variation is characterized as the standard deviation and used as the basis for comparison.

Two notes of caution are in order. The numerical results presented are dependent on the assumption of a normal distribution for both the manufacturing and measurement processes. As a consequence, any conclusions should be considered qualitative in nature. The distributions that may describe a real application must be investigated before the assumption of statistical normality can be invoked. If these distributions are not normal, complex statistical derivations become necessary and the numerical results derived here are no longer valid. Furthermore, the example cited above should not lead the dimensional measurement planner into complacency. Using state-of-the-art manufacturing equipment, it is quite probable that the process triangle may approximate an isosceles right triangle, where the legs are now of equal length. In such a case, the manufacturing process variation and the measurement process variation statistically combine such that there are nearly equal contributions from manufacturing and measurement to the overall process variation.

Returning to the example, Figure 8-2 shows the increase in observed process variation as a function of measurement process variation. The graph shows that measurement process variation must be approximately 45% of manufacturing process variation before a significant effect (i.e., greater than 10%) is encountered. The impact of 10% may or may not be an issue in the final accept–reject decision. Knowledge of the actual process distributions becomes important when determining the impact of the measurement process on production yields. If the manufacturing process distribution (the natural limits) falls well inside the specification limits, then the measurement process can have a relatively large amount of variation and still not significantly affect product acceptance decisions. How-

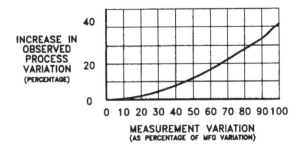

FIGURE 8-2 Effect of measurement variability on total process variability. Note: Assumes normal distribution. Intended to show approximate relationship only.

ever, if the manufacturing process has a distribution that just fits within the specification limits, then even a small amount of measurement variation will cause the product to fall outside the specification limits and cannot be tolerated.

Some obvious conclusions can be drawn:

1. Measurement planning requires knowledge of the natural limits of the manufacturing and measurement processes and corresponding product specification limits.
2. Measurement process variation increases the apparent variation of the complete process (i.e., manufacturing and measurement).
3. Selection of the measurement process must take place within the context of the specification limits and the manufacturing process capability.
4. Relatively speaking, very wide specification limits and a correspondingly tight distribution for the manufacturing process allow larger variation in the measurement process as a feasible choice. Conversely, narrow specification limits approaching the spread of the manufacturing distribution require a measurement method with small uncertainty.
5. Choosing one measurement process from all the feasible processes will be based on factors that include, but are not limited to, metrology.

8.2.2 Measurement Quality

Formal measurement planning eliminates the default planning that takes place on the production floor or in final inspection. This informal, undocumented, and uncontrolled planning is inevitably done by personnel who change with the shift, as does the measurement plan. With formal measurement planning, one presumes there is a planner with the requisite knowledge to do the job. This requires a trained and qualified metrologist capable of ensuring acceptable business decisions based on good metrology practice and sound economics.

Due to economic constraints, the cost of measurement is balanced against measurement quality. Inspection costs include the obvious task-related expenditures and also hidden costs associated with incorrect business decisions based on inappropriate measurement processes. This latter category of cost addresses the Type I and Type II errors defined in statistical theory. With the Type I error, the product that should have been accepted is rejected; with the Type II error, product that should have been rejected is accepted. Either of these errors can be reduced to a dollar value that reflects the cost of incorrect decisions. In this context, the concept of measurement quality becomes important.

Measurement quality is defined in terms of uncertainty. Uncertainty is tied to the ever-present variation resulting from one's inability—or reluctance due to business concerns—to completely describe what will be measured and how this will be done. The underlying issues and concerns are discussed in the previous chapter. This chapter examines how the uncertainty of the measurement process and the economics of the business situation can be integrated to generate acceptable measurement plans.

A measurement method is considered to be acceptable if it results in an acceptable measurement uncertainty. The general criterion for acceptability is an appropriate balance between measurement quality and cost. The decision is context-sensitive and is decided within overall business constraints.

One source of information on measurement planning is the ASME B89.7.2 standard. It contains supplemental material in its appendix that identifies uncertainty contributors and shows how to develop a measurement plan and select gaging. There is also an extensive exposition of the concept of uncertainty as it relates to pass–fail probabilities to be used to assess the acceptability of a measurement. In the case of these calculations, the statistical requirements include an understanding of the manufacturing and measurement process distribution.

The intent of measurement planning is to provide consistent guidance in the metrology area, balancing business needs with cost of inspection. The more rigorous and resource-intensive aspects of any measurement plan are subjected to scientific and business judgments to ensure that the ultimate business purpose is achieved and that the measurement process does not degenerate into an extravagant, scientific exercise.

8.2.3 Plan Content

The actual measurement plan must be linked to a specific product and process. The plan may provide measurements for product development, process monitoring, or product acceptance. Product characteristics subject to planning must have a functional purpose and relate to controllable manufacturing parameters.

The resources needed to create a plan include the availability of a dimensional measurement planner who has the knowledge and experience to carry out the planning tasks. The primary source of information the planner will use is the product specification, including drawings and other documents that give definition to the product. This is consistent with the view taken throughout this book. Even in small firms, it is expected that these duties will be explicitly assigned to an individual with the requisite training and authority to carry out the tasks.

Since the measurement process plan is linked with the manufacturing plan, the measurement planner must have knowledge of the manufacturing process characteristics. These include operational characteristics and failure modes. An understanding of the statistical distribution that underlies the manufacturing process is also required when the probability of pass–fail errors is to be used as one of the criteria in setting the measurement process.

The measurement plan should include at least the following items:

The key characteristics that are to be measured
Appropriate definitions of the characteristics measured
Complete specification of the measurement procedures
Environmental parameters that significantly influence the measurement process
Measurement strategy including operation sequence and sampling strategy
Methods of data acquisition
Data evaluation techniques
Target uncertainty

The general flow of measurement process planning assumes a target uncertainty has been established to verify the output of a specific manufacturing process. Once this target is identified, candidate gages can be selected and uncertainty contributors analyzed. With uncertainty for a particular gage or inspection procedure established, the measurement planner can calculate the pass–fail characteristics of the measurement method and estimate costs of inspection. This information provides the basis upon which the acceptability of the plan can be judged.

8.3 INSPECTION PROCESS UNCERTAINTY

Any measurement includes error—which, under most circumstances, may more correctly be referred to as uncertainty (Phillips, 1995). It may be caused by environmental changes, optical distortions, improper inspection equipment, inaccuracies in the equipment, incorrect procedures, or the distortions imposed on the part by the stresses of clamping it for measurement. Much of this uncertainty can be compensated for or minimized if recognized early in the design process. When a measurement is provided, the resulting information should include the value of the desired parameter and an assessment of the confidence in this value. The assessment can be provided by a statement of the uncertainty using the standard deviation as the basis or an explicit statement of a statistical confidence level.

Uncertainty values lead to the evolving concept of an uncertainty (or error) budget. Such a budget allows intelligent planning and design of the measurement system. Once an acceptable system uncertainty is determined, the budget allows apportioning this uncertainty to the elements of the measurement system.

Additional guidance in calculating uncertainty values can be found in ISO/TR 14253-2, which is a guide to the estimation of uncertainty in product verification. This document contains a more extensive list of uncertainty contributors along with worked examples. Table 8-1 contains an adaptation of the major categories of contributors found in the technical report. The remainder of the chapter discusses some of the uncertainty contributors from the list that are associated with a typical measurement encountered in product development. It is not exhaustive but is included to demonstrate general classes of uncertainty contributors that should be considered and estimates of magnitudes that may be encountered.

The numbers quoted in the subsequent discussion are from a variety of sources, both manufacturers' literature and other technical sources. The body of knowledge pertaining to dimensional measurement uncertainty is undergoing development as this book is being written. As a consequence, there are a variety of terms manufacturers and practitioners use that attempt to capture the intent of uncertainty analysis but do not provide the definitional precision required by standards. Many of these terms have historical legacies that are embedded in manufacturer specifications.

What follows is intended to provide the order of magnitude of uncertainty for selected dimensional measurements and to indicate the caution that must be taken when embarking on precision measurement tasks. An extensive table of

TABLE **8-1**   Uncertainty Contributors

Environment for measurement
Reference elements of equipment
Measurement equipment
Setup
Software and calculations
Metrologist and operator
Measurement object (workpiece)
Definition of characteristic
Measuring procedure
Physical constants

Source: ISO/TR 14253-2

characteristics and applications of measuring equipment can be found in Mil-HDBK-204A, including a column in the table identified as "accuracy." The accuracy values quoted most likely include some system effects that incorporate more than a single uncertainty contributor in the tabulated figure. Again, these estimates are intended to give a feel for the magnitudes of the errors (or uncertainties) inherent in any measurement process. The information required to meet the uncertainty requirements found in existing standards is more involved and requires process-specific information that could not be incorporated into a general table.

8.4 TOLERANCE CHARACTERISTICS AND MODELING

To verify whether or not a part conforms to its tolerance specification, the process planner needs a representation of the actual part. This representation or model is necessary when the infinite set of points that comprises the component becomes a finite data set in the inspection process. The inspector moves from the real part to a mathematical approximation by reducing the number of data points used for the description. Even if it were possible to reconstruct a perfect likeness of the part by inspection, economics require using a more limited set of points to build the model.

The inspection model is used to decide if the tolerance requirements have been met. This is accomplished by substituting extracted geometry—model geometry derived from the point data set using any number of techniques—for the actual feature on the part. The modeling techniques that may be used impose limitations on the inspection results. The model is then compared to the product definition and deviations are computed. These deviations represent the difference between the product description in the engineering drawings and the inspection model. The numeric values of the deviations are used to assess compliance.

To perform the comparison, the definition of the tolerance characteristic must include the underlying mathematics needed to obtain the characteristic's value. As may be suspected, these mathematical procedures introduce uncertainty. At least two distinct categories of uncertainty may be identified: (1) sampling uncertainty associated with the data points and (2) uncertainty introduced by the algorithm used to fit the substitute geometry. Sampling uncertainty includes sampling strategies involving the size of the data set, sampling density, and the influence of part geometry on sampling strategy.

Different values of uncertainty are introduced into the measurement by approximations that may be used to characterize the part's geometry. These approximations include substituting idealized geometry for the actual part features, deriving different representations of part surfaces using different modeling methods, and introducing correspondent substitute geometry (e.g., an axis) and tolerance characteristics. One can demonstrate how these uncertainty contributors enter into the measurement process by describing representative classes of models that may be used to characterize a feature and its tolerance.

Starting with the nominal geometry in the product definition [Figure 8-3(a)] and the fabricated component [Figure 8-3(b)], the inspector acquires a data set by traditional inspection techniques or coordinate metrology. The former methods may include (1) an extensive but not infinite data set utilized by a functional gage or (2) a more limited set created by surface-plate methods. Coordinate technology yields a cloud of points [Figure 8-3(c)] extracted from the actual

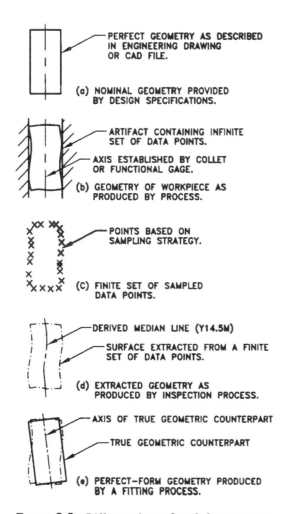

PERFECT GEOMETRY AS DESCRIBED IN ENGINEERING DRAWING OR CAD FILE.

(a) NOMINAL GEOMETRY PROVIDED BY DESIGN SPECIFICATIONS.

ARTIFACT CONTAINING INFINITE SET OF DATA POINTS.

AXIS ESTABLISHED BY COLLET OR FUNCTIONAL GAGE.

(b) GEOMETRY OF WORKPIECE AS PRODUCED BY PROCESS.

POINTS BASED ON SAMPLING STRATEGY.

(C) FINITE SET OF SAMPLED DATA POINTS.

DERIVED MEDIAN LINE (Y14.5M)

SURFACE EXTRACTED FROM A FINITE SET OF DATA POINTS.

(d) EXTRACTED GEOMETRY AS PRODUCED BY INSPECTION PROCESS.

AXIS OF TRUE GEOMETRIC COUNTERPART

TRUE GEOMETRIC COUNTERPART

(e) PERFECT—FORM GEOMETRY PRODUCED BY A FITTING PROCESS.

FIGURE 8-3   Different views of workplace geometry.

artifact that may be varied in both the numbers of points recorded and the effective density at various locations on the surface.

In the case of the functional gage [Figure 8-3(b)], the method of acquiring the data set precludes use of the individual points for analysis. The result is a Go–Not Go decision (i.e., an attribute gage). The gage lets the inspector determine whether or not the product will assemble but may not give sufficient accessibility to the information embedded in the data set to support downstream product development. For the functional gage, there is no opportunity to further refine the data and improve the model.

Where coordinate metrology is employed, a surface can be fitted to the recorded points [Figure 8-3(d)]. At this level many possible fitting techniques may be used, each altering the complexion of the resulting surface, each with its own characteristic effects on uncertainty. Furthermore, the density of the point data set (local and total densities) controls the surface generated by the fitting routine. The result is uncertainty concerning the boundary of the surface geometry and consequent uncertainty in the functionality of the component.

It is expected here that the fitted surface will contain deviations from the idealized geometry in the product description. This type of modeling is similar to fitting a styled Class I surface for industrial design purposes. For many inspection applications, this resultant surface may be more detailed than desired; the higher-resolution imperfections that are not needed for the intended purpose of the model vis-à-vis the product definition are not filtered out. Consequently, the model will not provide the desired information.

An alternative would be to fit idealized geometry to the part. This could be as simple as fitting a perfect cylinder to a turned part with the model's size, orientation, and location driven by the inspection data [Figure 8-3(e)]. The inherent uncertainty is illustrated by imagining a least-squares fit used to verify assembly of the component with mating parts. The underlying least-squares model would have point deviations both inside and outside the fitted cylinder. Since a maximum circumscribed cylinder might more accurately verify assembly, the points lying outside the least-squares cylinder may well interfere in the assembly operation, making functional verification uncertain.

The part feature is further idealized if the fitted surface is used to derive additional geometric elements that may be associated with the feature. The obvious example is an axis derived from the fitted geometry. The uncertainty in placing the axis in three-dimensional space is compounded by the uncertainty of the parent geometry and the fitting techniques used in arriving at the equation of the axis. The uncertainty propagates throughout the calculations as successive levels of geometry are created from the data set.

With the preceding concepts in mind and an uncertainty requirement for the measurement in hand, the usefulness of inspection results depends on the selection of a modeling method and the mathematical algorithms. The information contained in the data set must be transformed into a value that can be compared with the product specification. The fitting algorithms used by coordinate metrology equipment that produces different results depending on the choice of algorithm provide an example why the planner must understand the general classes of modeling and the purpose of the measurement.

8.5 SETUP

Before the metrologist can begin making a measurement, the part must be located with respect to a surface plate, an optical comparator, or CMM. The measurement equipment or fixtures establish the DRF. Parts are set up on inspection equipment so that those part datum features comprising the DRF are aligned with the simulated datum planes established by the equipment—which includes the virtual machine created by computer software in coordinate metrology. Measurements are then made from these planes.

In the subsequent discussion of measurement techniques, the underlying premise is that trends in metrology have moved away from surface-plate inspection toward the use of CMMs and similar equipment. This type of equipment incorporates greater metrological skill and ability in the machine. Through ignorance, this is transformed into a requirement for less capable operators, which is really not the case. While CMMs are certainly an attractive option yielding more flexibility and consistency along with the obvious increase in measurement productivity when applied correctly, it also raises planning issues. The major issue is whether these methods sufficiently emulate functional gaging results to support the desired range of product decisions and ensure assembly.

To simulate functional gaging with machines such as CMMs or optical comparators, the emphasis will be on using this equipment with fixtures to provide the correct DRF for measurement. In the event that fixtures are not incorporated into the process for economic or technical reasons, it is expected that computer-controlled CMMs or comparators will be used. The CNC equipment will ensure repeatable and reproducible DRFs, although not necessarily the ones established by a true functional gage. It is problematic as to which of these techniques (i.e., CMMs with or without fixtures) will yield the desired results. There are varying opinions on this matter, and there are no well-established methods for estimating uncertainties when using CMMs. However, manual probing will not establish the identical DRFs as fixtures or plate setups since this probing technique does not repeat with sufficient accuracy.

8.5.1 Datum Planes

The primary datum measurement plane is established by fastening the primary datum feature of the part at right angles to the principal datum feature simulator of the inspection equipment (surface plate, optical projector collimated light beam, or CMM axes). The secondary and tertiary datum planes are established by additional inspection fixture elements brought into contact with part datum features or by rotating the part 90° to bring first the secondary and then the tertiary part datum features into contact with the inspection surface.

The primary datum specified for a cylindrical feature is usually its axis. The coordinate planes that intersect at the axis are established when the part axis is placed parallel to and a certain distance from the inspection equipment plane. However, an axis, while derived from an actual part feature, is not a real feature. It must be determined in relation to a feature's surface. The secondary and tertiary datums are established by features that respectively prevent motion along the axis and by a feature that stops rotation about the axis.

8.5.2   Point Contact

Since all real part features are slightly imperfect, part datum features contact inspection fixtures at high points. The clamped part feature (primary datum) should make contact with the inspection surface at three or more points. In theory, the secondary and tertiary datum surfaces usually make two- and one-point contact with inspection equipment, respectively.

If the designer has specified a form tolerance for the datum features, the contact points will all fall within a zone acceptable for the datum plane. If the designer has specified datum targets, the contact points can be located—and relocated—at the same place each time the part is set up. Whenever possible, these datum target locations should coincide with tooling points. Datum targets are contacted by inspection equipment using parallel bars and tooling buttons or pins, as Figure 8-4 shows.

8.5.3   Axis Angularity

When positional tolerances are specified in a typical assembly, the tolerance zone of an axial part feature is actually a cylinder within which the axis of a part feature must lie (see Figure 8-5). The axis must be at some angular orientation, usually perpendicular, to the primary datum surface. It is essential to the part's function that the actual orientation is accounted for during inspection, because it affects the reported value of the feature's actual size. For example, the greater the deviation from the specified orientation a hole axis leans, the smaller a perfectly oriented mating part (pin or bolt) must be to fit the hole (see Figure 8-6). This concern is reflected in the Y14.5M and Y14.5.1M standards, where definitions include actual size and actual mating size. The latter standard contains explicit recognition of this orientation concern in Figure 8-6, which that illustrates the actual mating envelope and the actual mating envelope at basic orientation.

The metrologist can determine the angularity of a hole axis by taking mea-

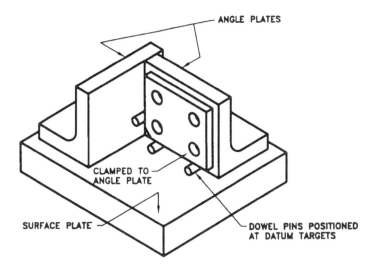

FIGURE 8-4   Setup using datum targets.

surements on both sides of the part over a short depth of the hole. However, this procedure does not tell the metrologist the full story about the circumferential surface of the hole. Inserting a close-fitting pin in the hole and taking inspection measurements at both ends of the pin yield more functional evaluation.

Tapped holes are not inspected inside the hole itself. A close-fitting, threaded pin is inserted and the measurements taken and recorded from the end of the pin that projects out of the hole, typically involving a projected tolerance-zone callout (see Figure 8-7).

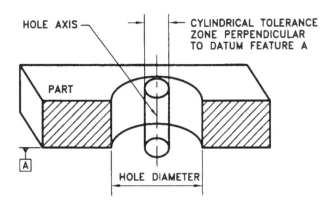

FIGURE 8-5   Theoretical positional tolerance zone.

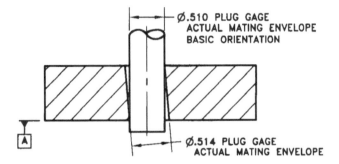

FIGURE 8-6 Actual size checks.

8.6 TEMPERATURE CHANGES

Temperature is considered one of the largest causes of uncertainty affecting measurements. Even under controlled conditions, different temperature gradients (transient and steady state) occur throughout a metrology laboratory, affecting both the part and the gaging device. Because all linear measurements are based on a controlled temperature of 68°F, or 20°C (as specified in ANSI and ISO standards), any measurement taken at a higher or lower temperature must be adjusted. This adjustment allows for the difference in expansion between the

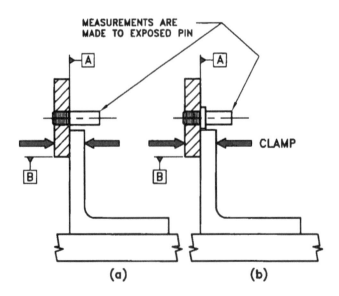

FIGURE 8-7 Inspecting tapped holes.

TABLE 8-2 Metal Expansion as a Function of
Temperature Change

Metal	Expansion from 68°F to 98°F (linear in. per in.)
Brass	0.00030
Aluminum	0.00040
Steel	0.00018
Stainless steel	0.00026

material in the measuring instrument and the material being measured. Table 8-2 shows expansion values for various metals at a 30°F increase in temperature. Predictions based on these values can vary up to 40% due to variations in heat treatment, internal structure, uncertainty in the coefficient of thermal expansion, etc.

8.7 EQUIPMENT INACCURACIES

Surface Plates. Even though open-setup inspection on the surface plate has been the final authority for acceptance—and the basis for gage calibration—it does involve inaccuracies. In many instances these plates are incorporated into the design of CMMs, so that their accuracy is still of concern even with the latest metrological technology. The slight variations of flatness and perpendicularity encountered in surface-plate equipment are lumped together and called "form error." These variations degrade inspection measurements. Furthermore, any indicator used in measuring is neither perfectly perpendicular nor perfectly parallel to the inspection surface, which may introduce cosine error into the readings.

Surface plates vary in size from approximately 8 by 12 in. up to 72 by 144 in. Plate thickness varies with plate size and can range from 1 to 24 in. Surface-plate flatness also varies from 0.00005 in. on the smallest plate up to 0.004 in. or more on the largest. These plates may be made from either granite or cast iron.

Working plates are divided into two accuracy grades. For one particular company, a grade *A* plate of average size (24 by 36 in.) is flat within 0.00017 in. A grade *B* plate of the same size is accurate to 0.00034 in. There is also a grade *AA* laboratory plate that is flat to 0.000085 in.

Surface-plate accessories—angle plates, parallels, V blocks, and sine bars (including those with centers)—are usually accurate within at least 0.0004 in. A *digital height gage* is accurate to 0.001 in. The height gage indicator used as a transfer device in measuring from a gage-block stack to a part is usually within 0.0004 in. *Hand micrometers* are accurate to 0.001 in. and certainly cannot measure reliably to 0.0001 in. Dial indicators are available in a variety of forms, capable of discriminating from 0.001 to 0.0001 in. In skilled measurement applications, they should be accurate within ±0.5 graduation. Indicators graduated to 0.001 in. usually have better repeatability characteristics than indicators graduated to 0.0001 in. because of less gearing, and therefore less backlash and hysteresis. They are also less sensitive.

The cited degrees of accuracy in surface-plate equipment can add up to a surprisingly large total uncertainty in some cases. As a consequence, many measurements discriminating to 0.0001 in. are obviously not valid unless the inspector has calibrated that portion of the surface plate he or she is using and each of his or her accessories. Additionally, great care is necessary to create a measurement process and environment that can resolve at this level.

Optical Comparators. The table and staging fixtures of an optical projector resemble surface-plate equipment in miniature and are used to align parts in essentially the same way. As a result, they are subject to the same form and orientation errors. Other sources of uncertainty unique to optical projectors lie in the optical characteristics and optical elements of these devices.

Magnification error in the optical projector consists of uneven magnification over the entire screen. Usually the center of the screen is most accurate, and the accuracy falls off toward the edges. Most optical projectors at 10 times magnification have a total magnification error over the screen of ±0.0003 in.

Chart gages (or templates) used with optical comparators are subject to various types of uncertainty. The lines on a chart gage vary in width from 0.004 to 0.006 in. as the instrument used in manufacturing them wears down, and their location may vary ±0.001 to ±0.005 in. Working to one side of a line can minimize width uncertainty. To reduce parallax, the line should be placed on the face of the chart toward the part shadow, not toward the metrologist. Drawings that define chart gages may be dimensioned so that chart gage lines are consistently located from the center to the inside or outside of their width.

Chart gages are usually made of plastic or glass. Mylar and vinyl are subject to distortion from temperature and humidity. With a 20°F change in temperature, the average size chart of Mylar will expand 0.008 to 0.010 in. per in., and vinyl 0.018 to 0.022 in. per in. A 20% increase in humidity will cause Mylar and vinyl to expand 0.010 in. per in.

Ordinary glass will expand and contract with changes in temperature but is unaffected by humidity. Moreover, glass has nearly the same temperature expansion and contraction rate as steel, so it is not an independent factor in the optical projector system. Obviously, glass is the most stable material to use in making chart gages and is therefore recommended. In any event, all chart gages should be calibrated so that their uncertainties are known.

The light source of an optical comparator is another cause of uncertainty.

The calibrated source itself is not essentially in error (except that a filament has area and therefore is not a point source of light), but it heats and changes the part, especially when reflected light is used to examine the part surface. Decreasing the amount of time the part is exposed to the light source can reduce this uncertainty.

Carefully used, optical projectors can give greater overall accuracy than a surface-plate check because dimensional relationships are easily visualized, no contact pressures are involved, and magnification errors are divided by the magnification factor.

Coordinate Measurement Machines. The table and probes of CMMs are not geometrically perfect in form or alignment. There are also scale and optical system uncertainties in a digital device. These machines must be calibrated before they are used. It should be noted that uncertainties in the design and fabrication of the basic machine are mapped and compensated for in the software of many pieces of equipment. Because CMMs use many types of probes, the effects of probe uncertainty must also be considered. This includes calibration of the probe along with the basic machine.

Another aspect of CMM usage involves the sampling routines. Since the probing of a component generates a discrete data set, uncertainty can be introduced by using an inappropriate sampling or inspection strategy. Particular emphasis needs to be placed on balancing accuracy with the desire to increase the inspection rate by taking the minimum number of points and using high positioning speeds. Techniques that may increase apparent productivity can markedly degrade the accuracy of the machine. The problems encountered in this area argue for careful planning and specification of the inspection process.

8.8   OPERATOR-INDUCED UNCERTAINTY

The metrologist can be a significant cause of uncertainty—through bias, faulty observation related to the nature of his or her equipment, or mistakes in setup and computation. In addition, there are issues related to education, training, dedication, and a host of nontechnical concerns.

8.8.1   Bias

Knowing that no two parts are exactly alike, anyone measuring the same part 10 times, but thinking that they are measuring 10 different parts, will arrive at a series of 10 different measurement readings, varying as much as 0.001 in. This has been experimentally verified many times.

Another form of bias occurs in relation to the source of parts. A metrologist checking parts coming from a supplier is inclined to be more severe; he or she will reject marginal parts. If his or her own company has manufactured the outgoing parts, the metrologist will be inclined to pass parts close to or at the tolerance limits.

A third form of bias has to do with the metrologist's idiosyncrasies in using measuring instruments. For example, he or she might tighten up a bit too much with a micrometer, causing it to act more like a clamp. In this case, the reading will be smaller than the part actually justifies. Even the movements involved in comparing a part to a gage block stack with a height gage and indicator are subject to inspector error.

8.8.2   Observation

The parallax factor is a constant source of observational uncertainty. Neither can an indicator needle rest directly against the markings on a dial face, nor can the scribed lines on a chart gage be exactly at the focal plane of an optical projector. Even the barrel markings on a micrometer can be misaligned because of parallax. Digital instruments are certainly helpful in reducing parallax and bias. However, even with these instruments, the digital readout may be blinking back and forth between adjacent values, forcing a choice to be made.

8.8.3   Computation

If computation is required, the metrologist is likely to make mistakes now and then, even in simple addition and subtraction. Some of the computational error can be reduced through the use of direct digital data acquisition, as the raw data will be recorded in exact form.

8.8.4   Setup

Setup uncertainty can occur even though the designer has clearly specified the necessary datums to ensure proper orientation and location of the part. The metrologist might

1. Use the wrong datums
2. Contact the part at other than the prescribed datum target locations
3. Not get the primary datum surface of the part perpendicular to the collimated light beams of the optical projector
4. Not focus the part properly in the optical projector, which could distort the part shadow and throw off direct measurements or comparison of the part shadow with optical chart gage lines

8.9   FREE-STATE VARIATION

When part tolerances are less than 0.002 in., clamping a part for inspection causes perhaps the most significant percentage of uncertainty introduced into inspection measurements. Clamping stress distorts the part, changing its form and dimensions slightly. All parts are flexible, some to a greater degree than others are.

The part should, if possible, be restrained during inspection to the same degree it will be at assembly. This can be included in the product definition, possibly allowing a larger tolerance in the free state with a more stringent tolerance under the conditions of restraint. The restraining conditions would be specified through the use of a note. The part is "frozen" in only one orientation to the datum reference frame, reducing what could be extreme uncertainty in the measurement process.

With complete specification of the component's state during the measurement process, metrologists would no longer be allowed to compensate for variation in flexible components by unclamping a part and adjusting it. Repeatedly releasing the part and then clamping and measuring it again in a quest to find the one position that allows the part to meet acceptance are time-consuming and not a reliable measurement method.

8.10 RECORDING INSPECTION RESULTS

For measurements to be useful, they must be recorded in an understandable format as they are made. An organized and easily read format will allow quick evaluation by engineers, designers, and inspectors to verify acceptance or rejection of a part or batch of parts. A standard, detailed report form would be very helpful for complete recording and meaningful evaluation (see Table 8-3). It would also help clear up the many misunderstandings and ambiguities that commonly occur during engineering analysis and in material or manufacturing review board meetings.

8.10.1 Recording Setup

One of the chief objections to most measurement reports is a lack of provision for recording part setup. If a datum reference frame has been specified on the drawing, it is of paramount importance that the metrologist indicates which datum surfaces used in setup so that they can be verified during engineering analysis. Ideally, the functional datums specified in the component definition are used for inspection.

The metrologist should also specify the nature of any setup devices to be used to simulate specified datum targets. Two dowel pins might orient a part differently than a parallel bar, for example, especially if the datum surface is irregular. Keep in mind that the product definition specifies the simulated datum features found in the setup. The dowels and parallels just mentioned do not provide identical setup methods. The preferred method of specifying the setup is

TABLE 8-3 Sample Inspection Report

Inspection Report

Figure: Datum; setup: Surface A and two additional sides. Datum targets used are marked on part.

	Specified dimension, tolerance, etc.				Actual dimension as checked					
Item no.	x	y	Tolerance diameter at MMC	Size	x	y	Tolerance (actual)	Tolerance diameter allowed by MMC	Size	Results
1	0.520	2.020 (0.520 + 1.500)	0.000	0.500 0.540	0.538 0.540(A)	2.033 2.030(A)	See paper gage[a]	0.030	0.530	OK
2	2.520 (0.520 + 2.000)	2.020 (0.520 + 1.500)	0.000	0.500 0.540	2.548 2.545(A)	2.000 2.000(A)	See paper gage	0.035	0.535	OK
3	2.520 (0.520 + 2.000)	2.020 (0.520 + 1.500)	0.000	0.500 0.540	2.522 2.522(A)	0.497 0.497(A)	See paper gage	0.020	0.520	OK
4	0.520	0.520	0.000	0.500 0.540	0.505 0.510(A)	0.525 0.525(A)	See paper gage	0.025	0.525	OK

[a] A paper gage for this inspection report is developed in Chapter 11 and would be attached to this report.

through the use of process sheets calling for methods identical in form and function to those found in the fabrication process and in functional use.

If no inspection process sheet exists and no space is available on the measurement report, a most convenient and informative procedure to record the setup is to take a snapshot using a camera or record the setup using a video camera and attach it to the inspection report. Another feasible alternative is the use of digital photography, which allows maintaining a computer database of setup information.

8.10.2   Recording Hole Axis Angularity

Another common deficiency on inspection reports that provide positional information is the lack of data on axis angularity. This information is essential to functional acceptance, particularly if paper layout gaging is to be used to evaluate inspection data. In particular, note that the positional tolerance zone can act in a similar fashion to the zone that occurs when an orientation control is called out. Any interpretation of the positional results may need added information that allows partitioning the positional variation into both location and orientation elements. This would allow more intelligent decisions regarding corrective action. Thus, two sets of measurements should be recorded for each hole center.

8.10.3   Recording Tolerances

It is important that dimensions, tolerances, and inspection measurements be recorded on the report so that proper analysis can be made. In the case of a bilateral tolerance or an RFS callout, the tolerance allowance is a fixed requirement; however, an MMC callout, with or without specifying a zero tolerance at MMC, permits the positional tolerance to increase as the feature size departs from MMC. Each set of measurements may therefore involve a different positional tolerance (based on different actual size measurements), in which case the inspector must judge acceptance on an individual feature basis and record the actual tolerance allowed for each part feature.

8.11   CONCLUSION

A good criterion by which to judge the completeness of a measurement process is to ask, "Does it contain enough specific information to completely reconstruct the geometry of the part under inspection?" An affirmative answer is a prerequisite to meaningful engineering analysis, particularly if that analysis includes computer-simulated gaging (soft gaging) to establish functional acceptance of the part.

REFERENCES

ASME, B89.7.2, Dimensional Measurement Planning (Draft), New York: ASME, 1999.

ASME, Y14.5M-1994, Dimensioning and Tolerancing, New York: ASME, 1994a.

ASME, Y14.5.1M-1994, Mathematical Definition of Dimensioning and Tolerancing Principles, New York: ASME, 1994b.

Department of the Army, Mil-HDBK-204A(AR), Design of Inspection Equipment for Dimensional Characteristics, NJ: Dept. of the Army, 1990.

Haugen, E., Probabilistic Approaches to Design, New York: Wiley, 1968.

ISO/TS/4253-2:1999, Geometrical Product Specifications (GPS)—Inspection by Measurement of Workpieces and Measuring Equipment—Part 2: Guide to the Estimation of Uncertainty in GPS Measurement, in Calibration of Measuring Equipment and in Product Verification.

Lorincz, J., Finding the pluses in high-precision accuracy, Tooling & Production, May, pp. 29–33, 1993.

Nilsson, J. T., Application considerations, in Coordinate Measuring Machines and Systems, J. A., Bosch, ed., pp. 306–308, New York: Marcel Dekker, 1995.

Phillips, S. D. Performance evaluations, in Coordinate Measuring Machines and Systems, J. A. Bosch, ed., pp. 137–139, New York: Marcel Dekker, 1995.

9

Functional Gaging

9.1 INTRODUCTION

Preceding chapters present the conceptual foundation for integrated product development. The material shows how to start with a set of general, yet powerful, geometric and organizational concepts and use them to define a product and its allied processes. The use of geometric dimensioning and tolerancing with its emphasis on well-defined datums and geometric controls forces the design team to think ever more carefully and to define many geometrical characteristics not previously considered. The intentional application of geometric controls gives the product development team jurisdiction over final product quality by dictating the specific gage design criteria that determine functionally acceptable parts.

Once this geometric product description is in place, the process of designing the functional gage is used to validate the design definition and to verify conformance. As a corollary to this, the gage's resulting design also provides geometric form to the corresponding production tooling.

This chapter's material is limited to functional gages; that is, gages that "receive" the part being inspected and that contain fixed elements (pins, bushings, etc.) to check part features. These gages function as if they were mating parts. These functional or receiver gages simulate the most critical conformation of the mating part when they "receive" the part being gaged. They do not report variables inspection data but only tell the user if the part is to be accepted or rejected.

159

Two types of functional gages are discussed: (1) feature relation and (2) feature location and relation. Such instruments have fixed configurations, like the mating parts they simulate. The gage will allow each part being checked a different combination of size and geometric tolerances since no two fabricated parts can ever be identical. In the context of the following examples, the feature relation gage checks only the relationship of a pattern of part features. Such a gage contacts only the primary datum surface of a part. The feature location and relation gage checks the location of a part feature, or a pattern of part features, relative to a datum reference frame. Such a gage contacts two or more part datum features. Nonvariable tolerancing methods, such as bilateral tolerancing or positional tolerancing when not modified with the maximum material condition (MMC) callout, are not compatible with the fixed-configuration gages.

A functional gage is comprised of two classes of components: the fixture and the gaging elements. The fixture serves to locate the part and presents it for gaging. The gaging elements verify that the controlled features conform to the product definition. These features are specifically identified in the product specification as requiring verification.

The fixture (whether for gaging or tooling) can incorporate any of three distinct functions. As mentioned, its primary task is to locate the component in a manner that presents the part to the gaging (or inspection) elements, duplicating the orientation and location experienced in the actual assembly. All fixtures must accomplish this; otherwise the results of the inspection process will not yield information that can be directly compared to the product specification.

The fixture also may provide support to prevent uncontrolled deflection due to the part's weight or forces generated by the gaging or measurement process. When introducing separate supports that are distinct from the datum features simulators incorporated into the fixture, carefully ensure that the workpiece is positioned by the locators (datum feature simulators or targets) and not the supports. Whether fixed or adjustable, the supports must not serve as alternate locators. When they assist in actually locating the part, they prevent accurate location (i.e., duplicating the location provided by the functional DRF) of the part by the fixture and introduce uncertainty into the production and inspection processes. These effects must be considered in any design, because production personnel may need higher skill levels to compensate for lack of positioning accuracy.

The other function incorporated into the fixture involves clamping of the part. This may provide restraint during measurement or inspection if such a requirement is included in the product definition. It might also be needed if the part is oriented for inspection in a position that requires the fixture to introduce equilibrium forces to counteract gravity.

Illustrations have been developed to directly show gaging principles. These demonstrate the use of GD&T for dimensioning and tolerancing and show how it leads to simple and practically designed gages and tools. Functional gage design is an important element in achieving physical production due to its influence on the manufacturing sequence (process plan) and production tooling design. The remainder of this chapter focuses on the design of the gaging elements; further explanation of the gage fixture is discussed with workholding issues in Chapter 12.

9.2   FUNCTIONAL GAGING PRINCIPLES

A group of fundamental gaging principles is directly applicable to the design process if the structured geometric approach is followed. These principles are listed below. Keep in mind that these are general rules that need to be applied with thought and discretion. The physical aspects of the product's design and manufacturing processes determine how to implement the principles in a specific situation.

1. Gages, production tooling, and parts (all of which may include tolerances and wear allowances) should be designed simultaneously using a concurrent engineering team.
2. The gage designer should not have to make arbitrary decisions regarding gage element size, geometric characteristic, or location. A complete product specification dictates the gage design and interchangeability criteria through the use of appropriate geometric controls.
3. Gages should be defined with the same geometric characteristics used on the part being gaged.
4. The gages have companion features (with respect to the part) that provide the datum feature simulators incorporated within the gage. These datum feature simulators represent features on the mating component.
5. Functional gages have fixed gaging elements located at basic dimensions conforming to feature locations described in the product definition.
6. These gages may simulate (1) the worst-case (virtual condition) part if there is no fit allowance or (2) the worst-case (virtual condition) mating part if there is a fit allowance.
7. One datum reference frame (DRF) per part enables one gage to be used for acceptance. Any increase in the number of DRFs increases the number of gage and measurement setups.
8. All functional gage elements should "go" into or over the part features simultaneously where simultaneous requirements are invoked by the product specification.

9. A conscious decision should be made to establish gaging policy, weighing both producer's and consumer's risk. This decision may be set by contract or by reference to appropriate standards.
10. Parts that can be practically gaged can also be practically tooled since tools and gages should be interchangeable.

9.3 FEATURE RELATION GAGES

The gage element designs discussed in this section apply to all patterns containing two or more features, provided none of the features is designated as a part datum. Only feature relation gages are discussed; combined feature location-and-relation gages are covered in a subsequent section. All the gages discussed in this section must contact the primary datum surface of the part during use, and suitable mechanical means must be used to ensure contact. The following techniques are particularly useful for sheetmetal and other similar types of flexible parts where the illustrated feature patterns may, as a group, be used as datum features of size.

9.3.1 Internal Feature Patterns

9.3.1.1 Clearance Hole Patterns

Figure 9-1(a) shows a six-hole pattern of $\varnothing.510-.530$-in. clearance holes in which each hole has a positional tolerance of $\varnothing.010$ in. when the hole is at MMC. Note that only the primary datum A on the part is specified, which indicates that just a feature relation gage is required. This gage checks the hole pattern relationship and perpendicularity to datum plane A, but not the location of the pattern on the part.

Functional Gage Design.   The gage element locations are identical to the basic dimensions on the part drawing (both patterns are identical), and gage element pin G [Figure 9-1(b) or (c)] must have a minimum height that at least equals the maximum thickness of the mating part. This gage element pin checks hole location and perpendicularity to datum plane A, and the six gage element pins are $\varnothing.500$ in., determined as follows:

$\varnothing.510$ in., hole H at MMC

$-\ \varnothing\underline{.010}$ in., positional tolerance specified at MMC

$\varnothing.500$ in., basic size gage pin G

All six gage element pins must enter the six part features at the same time, and the gage datum element must contact part datum surface A for acceptance. Thus, H minus G equals 0.010 in. at MMC.

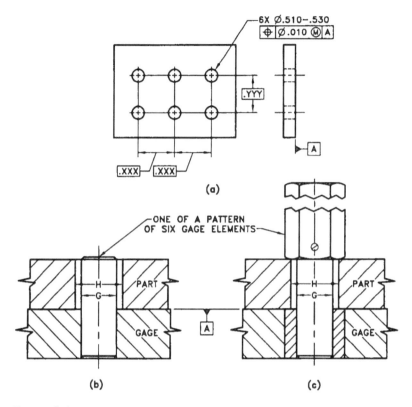

FIGURE 9-1   Part with clearance holes and functional gage elements.

Plug Gages Required.   A ⌀.510-in. Go gage and a ⌀.530-in. Not Go gage are required. If the drawing callout required ⌀.500–.530 (6 places) with positional tolerance ⌀.000 in. at MMC, the identical ⌀.500-in. gage elements shown in Figure 9-1(b) or (c) would still be required (⌀.500-in. hole H minus ⌀.000-in. positional tolerance equals ⌀.500-in. gage element G). The ⌀.510-in. Go plug gage would not be required, however, because it is built into the functional gage in the form of the six ⌀.500-in. gage elements that simulate the six ⌀.500-in. bolts that will fasten this part to the mating part. Figure 9-1(c) illustrates another functional gage with separate gage elements G inserted into the part holes and then into close-fitting (0.0001–0.0002-in. allowance) bushings in the gage base. The same basic rules apply to this design as to the gage shown in Figure 9-1(b), and all six gage elements must enter the bushing and be in place in order to accept the six-hole pattern for perpendicularity and feature relationship.

Specific applications of the gage elements illustrated in Figure 9-1(b) or (c) are shown in subsequent examples, but it can be generally stated that the separate gage elements shown in Figure 9-1(c) will enable the gage operator to determine which of the six part features is out of tolerance. All six gage elements must enter the six holes in the part and be in place simultaneously. The number of separate gaging elements [as shown in Figure 9-1(c)] would never be reduced for checking any feature pattern by "pinning" several holes and then "walking" one gage element around the part pattern. The part–gage relationship could easily shift during this type of gaging operation, and out-of-tolerance parts could be erroneously accepted. The entire basic feature pattern must be gaged simultaneously, as all bolts must be inserted as a pattern at assembly.

9.3.1.2 Tapped Feature Patterns

Figure 9-2(a) shows a six-feature pattern of tapped holes. Figure 9-2(b) shows the basic design of the gage that inspects each tapped hole in the pattern. Gage element G, which goes through bushing B in the gage base and enters the tapped hole in the part, simulates the bolt, and all six of these gage elements must be in place at the same time.

Functional Gage Design.   The bushing locations on the gage base are identical to the basic dimensions on the part drawing (both patterns are identical), and the gage bushings have a minimum height that at least equals the maximum thickness P of the mating part (0.50 in. in this example). The difference between gage base bushing diameter B and that portion of the Go thread gage element G (that simulates a bolt where it goes through the gage bushing) is the positional tolerance specified at MMC for the tapped features. Thus, B minus G equals 0.010 in. in Figure 9-2(b) or 9-2(c).

Plug Gages Required.   Gage element G is a Go thread gage, and so only a Not Go thread gage (not shown) is required to check each hole separately. Figure 9-2(c) shows the same basic gage design as in Figure 9-2(b) except that Go thread gage member G has been "stepped" so that a standard size bushing B may be used. In Figure 9-2(b), bushing B would have to be enlarged to obtain the 0.010-in. difference between the 0.500-in. G gage element, because a Ø.510-in. bushing is not standard.

Miscellaneous Considerations.   The same gage would check a "Helicoil"-type threaded insert, but threaded inserts with locking features should not be checked, because the Go thread element could damage the locking feature. It is better to gage the tapped hole pattern with a special in-process gage before installing the insert.

Because there can be a tolerance allowing 0.000- to 0.004-in. shake or movement between the pitch diameters of the Go thread gage elements and part

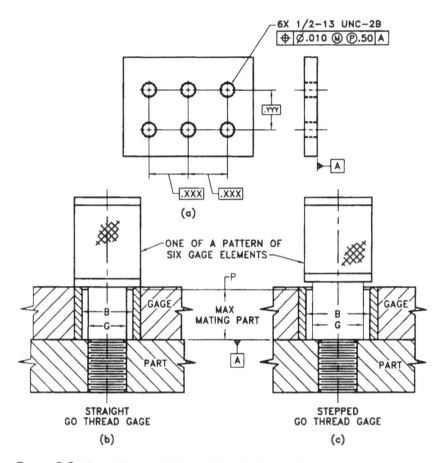

FIGURE 9-2   Part with tapped holes and functional gage elements.

tapped features, the MMC modifier is applicable. If the positional tolerance for the tapped hole were modified with "regardless of feature size," the Go thread gage element would have to be centered in the tapped thread. This would require either a tapered Go thread element or an expanding split-thread arrangement. Because very few mating parts have centering threads, the RFS callout is usually not practical.

Tapped features are gaged for relation and perpendicularity where the assembled bolt (simulated with the Go thread gage element) goes through the mating part feature (simulated with the gage bushing). The tapped thread is not checked for relation or perpendicularity inside the thread. Because the mating part thickness is quite critical in determining the allowable perpendicularity of the thread, this information is vital to the gage designer in determining gage bushing height and should be specified on the part drawing using the projected tolerance-zone symbol.

9.3.1.3 Counterbore Patterns

Figure 9-3(a) shows a part with clearance holes and counterbored holes. Figure 9-3(b) shows the basic design of the gage that inspects a pattern of counterbores and holes when both counterbores and holes have the same specified positional tolerance (i.e., each counterbore is not located from its respective clearance hole).

The bushing locations on the gage base are identical to the basic dimensions on the part drawing (both patterns are identical). The combined counterbore–clearance hole gage element must fully enter both counterbore and gage base bushing, and bottom on the counterbore. The difference between part counterbore C' and gage element C, and part clearance hole H' and gage element H, equals the Ø.010-in. positional tolerance specified when both counterbores and clearance holes are at their MMC size. Thus,

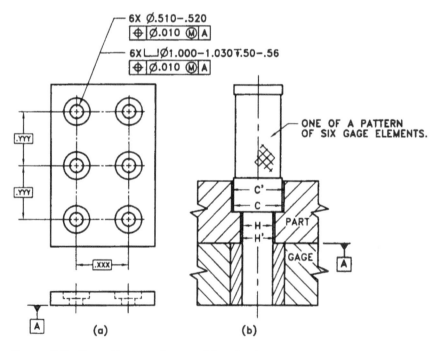

FIGURE 9-3 Part with counterbores and functional gage.

Ø.510 in., hole H′ at MMC

− Ø.010 in., positional tolerance specified at MMC

Ø.500 in., gage element H

For the counterbore,

Ø1.000 in., counterbore C′ at MMC

− Ø .010 in., positional tolerance specified at MMC

Ø .990 in., gage element C

Plug Gages Required. The following gages are required: (1) Ø.510-in. Go gage; (2) Ø.520-in. Not Go gage; (3) Ø1.000-in. Go gage; and (4) Ø1.030-in. Not Go gage.

9.3.1.4   Fixed-Nut Retainer Patterns

Figure 9-4(a) shows a pattern of fixed-nut retainers while Figure 9-4(b) shows the basic design of a gage that inspects a part containing a pattern of fixed-nut

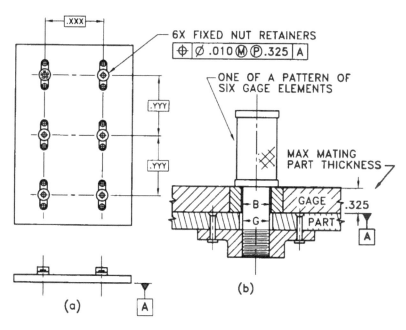

FIGURE 9-4   Part with fixed-nut retainers and functional gage.

retainers (i.e., the nut is rigidly held in the retainer). With such a pattern, only the location of the pitch diameter of the fixed nut in the retainer must be gaged. The access hole in the part need not be gaged for location. As long as the access hole allows gage element G (which simulates the bolt) to enter the fixed nut, it is large enough and properly located. The minimum gage bushing height (0.325 in.) shown in Figure 9-4(b) is the maximum part thickness (projected tolerance zone). The difference between gage base bushing diameter B and that portion of gage element diameter G (that simulates the bolt) is the positional tolerance specified for the fixed nut in the nut retainer. Thus, B minus G equals 0.010 in.

This gage is similar to the gage in Figure 9-2(b) as the fixed-nut retainer simply allows the designer to place a nut in an inaccessible location or on a part that is too thin to be tapped. The Go thread gage G element can be stepped, as shown in Figure 9-2(c). When designing the amount of "step," be sure that the Go thread gage element does not contact the access hole in the part.

9.3.1.5 Floating-Nut Retainer Patterns

Figures 9-5(b) and (c) illustrate the individual gages required to inspect the floating-nut retainers shown in Figure 9-5(a). The pattern relationship of the clearance

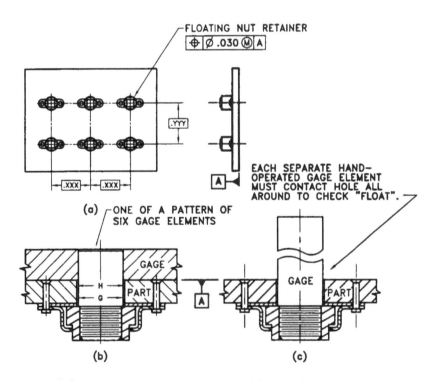

FIGURE 9-5 Part with floating-nut retainers and functional gage.

holes in the part (not the floating nuts) is gaged with the gage shown in Figure 9-5(b). This gage is a pattern of fixed-gage elements that must simultaneously enter their respective clearance holes. The fixed-gage pins are located at the basic clearance hole locations shown on the part drawing. Thus, H minus G equals 0.030 in., the positional tolerance at MMC. The separate handheld gage shown in Figure 9-5(c) is inserted into each floating nut to determine if the nut has sufficient "float" to allow this handheld gage (which simulates the bolt) to contact the clearance hole all around its circumference.

The mating part thickness is not specified in Figure 9-5(a) because the floating nut in the nut retainer can be tilted and thus adjusted for misalignment when the bolt [simulated with the hand held gage element in Figure 9-5(c)] is inserted. The fixed-gage elements mentioned above could fit into bushings in the gage base and be sequentially inserted into the part clearance holes until all are assembled [Figure 9-5(c)].

9.3.2 External Feature Patterns

These gaging principles apply to all patterns containing two or more features.

9.3.2.1 Stud Patterns

Figure 9-6(a) shows a six-stud pattern consisting of ∅.499–.500-in. studs welded on a plate. The weld flash has been removed. Figure 9-6(b) shows the gage, and details one of six bushings, located at the same basic dimension as the studs in Figure 9-6(a). The bushing internal diameter is determined as follows:

∅.500 in., specified MMC of the stud

+ ∅.010 in., positional tolerance specified at MMC

∅.510 in., bushing internal diameter B

Thus, B minus S equals 0.010 in. at MMC.

For acceptance, all six bushings must go over the six studs simultaneously, and the gage datum element must contact part datum feature A to accept the part. The ring gages required are a Go gage of ∅.500 in. and a Not Go gage of ∅.499 in.

9.3.2.2 Dowel Pin Patterns

Figure 9-7 shows a pattern of six ∅.4998–.5000-in. dowel pins in a plate. Figure 9-6(c) shows the gage, and details one of six bushings, located at the same basic dimension as the dowels in Figure 9-7. This gage is identical to the gage shown in Figure 9-6(b) (except perhaps for bushing tolerance), and the bushing internal diameter is determined as follows:

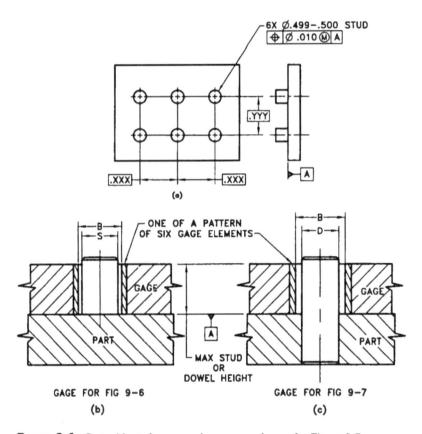

FIGURE 9-6 Part with studs, companion gage, and gage for Figure 9-7.

      Ø.5000 in., specified MMC of the dowel pin

+ Ø.0100 in., positional tolerance specified at MMC

      Ø.5100 in., bushing internal diameter B

Thus, B minus D equals 0.010 in. at MMC.

    All six bushings must go over the six dowels simultaneously, and the gage datum element must contact part datum feature A to accept the part. The ring gages required are a Go gage of Ø.5000 in. and a Not Go gage of Ø.4998 in.

9.3.2.3 Threaded Stud Patterns

Figure 9-8(a) illustrates a pattern of six threaded studs, and Figure 9-8(b) shows the gage (one of six gaging elements) that gages each stud in the six-stud pattern. The difference between the outer diameter G of the internal Go thread gage and

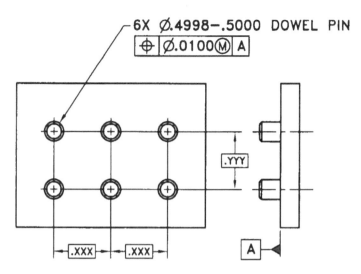

FIGURE 9-7 Part with dowel pin, companion gage shown in Figure 9-6(c).

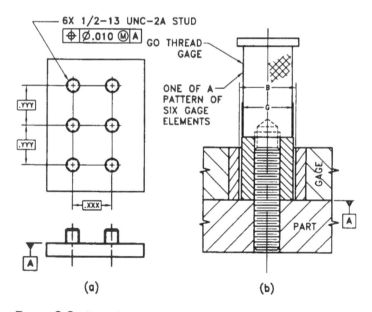

FIGURE 9-8 Part with threaded studs and functional gage.

the inner diameter B of the gage bushing is 0.010 in. MMC has been specified for the location of the threaded studs; this is, in effect, the "shake" possible between the Go thread gage and the stud threads. There can be from 0.000- to 0.004-in. "shake" in the Go thread gage in many cases, so the MMC modifier is applicable. If the positional tolerance for the tapped stud were modified with RFS, the thread gage element would have to be centered on the stud, and this would demand an adjustable split Go thread element.

In this instance, only positional and perpendicularity tolerances are being checked, based on the assumption that thread geometry has been previously checked and is acceptable.

9.4 DESIGN PRINCIPLES FOR FEATURE LOCATION AND RELATION GAGING

9.4.1 Critical (RFS) Part Datum Features

All the gages discussed in this section must contact the designated part datum features during use, and suitable mechanical means must be used to ensure contact.

Part datum features modified with RFS require gage-centering devices so the gage will center on the part datum feature, regardless of its finished size. This particular application does not meet the more restrictive definition of a functional gage as a gage having fixed elements. Yet it provides information in a similar fashion to the true fixed-gage design. Furthermore, it is a useful example to demonstrate the complications that arise when RFS applications are used with no functional basis.

Figure 9-9(a) shows a part datum feature modified with the RFS material condition and indicates, by the use of four tool and gage pickup points (datum targets), where the part will be "centered." The datum features modified with a RFS callout must, of course, be in tolerance.

Figure 9-9(b) shows the gage that meets this drawing requirement. The four dial indicators located adjacent to four of the holes in the pattern are specified with the datum target point symbols in Figure 9-9(a). The gage will have to be set with a master so that the indicators are zeroed (or so that the diametrically opposed indicators have the same readings) and, consequently, the part is centered when it is placed in the gage. Note that all eight holes are checked simultaneously as a pattern with eight fixed-gage elements located at the basic drawing dimensions. If the part assembles on the eight gage elements when contacting the gage on datum surface A, and then can be centered in the gage by zeroing out the indicators, the part will be accepted. Thus, the eight holes were located within their respective positional tolerances from that "center" specified on the part.

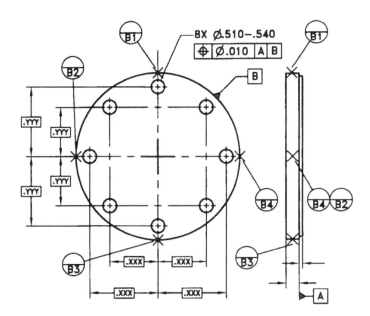

FIGURE 9-9(a) Critical (RFS) datum feature, with tool and gage pickup points.

Fixed Gage Elements. Each of the eight gage pins will be ⌀.500 in., determined as follows:

⌀.510 in., MMC diameter of the part holes

− ⌀.010 in., positional tolerance specified at MMC

⌀.500 in., basic gage pin size

A ⌀.510-in. Go plug gage and ⌀.540-in. Not Go plug are required.

Figure 9-10(a) shows a small central part datum feature modified with the RFS material condition. The part must freely pass over the ⌀.406-in. Go gage pin, and part datum surface A must then fit flat against the gage datum element [Figure 9-10(b)]. This step checks that the ⌀.406–.408-in. part datum feature is perpendicular to the datum planes at MMC.

Four steps are required in using the gage in Figure 9-10(c):

1. Locate part datum feature A on the gage base datum plane A.
2. Insert datum gage element with diameters D (⌀.406 in.) and D' (somewhat larger than ⌀.408 in.), and center the part datum hole on the taper between D and D'.

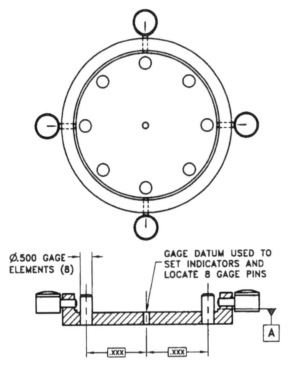

GAGE DATUM USED TO
SET INDICATORS AND
LOCATE 8 GAGE PINS

⌀.500 GAGE
ELEMENTS (8)

A

FIGURE 9-9(b)   Functional gage (with dial indicators) for a critical (RFS) datum feature.

3. Insert the G gage elements and rotate the part as necessary to effect their entry.

4. Position all four G gage elements at the same time to accept the part.

Note that in step 2, the use of the taper introduces some error since the entire bore is not utilized to determine the datum axis.

For this operation, three plug gages are required: a ⌀.380-in. Go gage; a ⌀.388-in. Not Go gage; and a ⌀.408-in. Not Go gage for the "datum" hole. The gage element in Figure 9-10(b) is the ⌀.406-in. Go gage.

Part datum features modified with the RFS material condition are quite difficult to gage and are encountered only when mating part datum features consist of a tapered centering device (or a force fit as an alternative) similar to that shown in Figure 9-10(c). It is very difficult to repeatedly pick up the exact "center" of a part datum feature unless that feature has perfect form.

Figure 9-11(a) shows tapered part datum feature A modified with the RFS specification. The clearance and counterbored holes and the two tapped holes are all dimensioned from the "center axis" of the taper. The mating part that contains

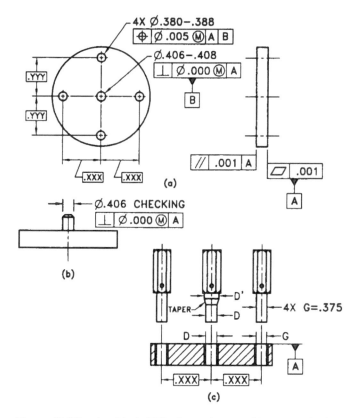

FIGURE **9-10**   A critical (RFS) datum feature with two required gages.

clearance holes for the two screws that enter the tapped holes is 1.00 in. thick, which the 1.00-in. projected tolerance zone indicates. Figure 9-11(b) shows a gage that verifies this part requirement. The tapered gage datum element should be the nominal size of the part datum feature.

The gage should be used as follows:

1. Insert the part into the gage and rotate if necessary so that the two Go thread gage elements enter the threads, and then screw them partially in place. Make sure that the part datum feature is firmly in contact with the gage datum element before and after tightening the two Go thread gage elements.

2. Insert the counterbore–clearance hole gage. H should be ∅.437 in., derived as follows:

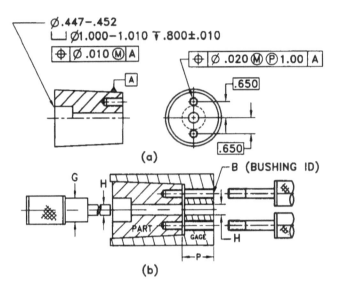

FIGURE 9-11 A critical (RFS) tapered-part datum feature and functional gage.

Ø.447 in., at MMC

− Ø.010 in., positional tolerance specified at MMC

Ø.437 in., H basic

G should be Ø0.990 in., derived as follows:

Ø1.000 in.

− Ø .010 in., positional tolerance specified at MMC

Ø .990 in., G basic diameter

3. For acceptance, all three gage elements should go into the part at the same time.

Three plug gages are also required: a Ø.447-in. Go gage; a Ø.452-in. Not Go gage; and a Not Go thread gage. The two thread gages that enter bushing B are Go gages.

9.4.2 Critical (MMC) Part Datum Features

Figure 9-12(a) shows a part that has MMC specified for the Ø.501−.502-in. part datum feature. This datum feature has been allowed a perpendicularity tolerance of 0.001 in. (at MMC) from datum plane A (which must be identified on symmetrical parts).

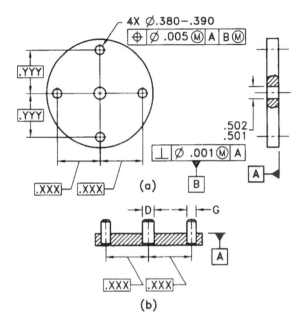

FIGURE 9-12 A less critical (MMC) single datum feature and functional gage.

The functional gage [Figure 9-12(b)] for this part must take into account both the MMC size of the part datum feature and the perpendicularity tolerance allowed on that feature by using a ∅.500-in. pin D as the fixed datum gage pin. This is determined by subtracting the 0.001-in. perpendicularity tolerance from the MMC size (∅.501 in.) of the part feature. The final gage is shown in Figure 9-12(b), and the four fixed-clearance hole gage elements will be

∅.380-in., holes at MMC

− ∅.005-in., positional tolerance when holes are at MMC

∅.375-in., G gage pins (4 required)

For this operation, four plug gages are required: a ∅.501-in. Go gage; a ∅.502-in. Not Go gage (for datum B); a ∅.380-in. Go gage; and a ∅.390-in. Not Go gage for the clearance holes.

If the part drawing did not specify a 0.001-in. perpendicularity tolerance for the part datum feature B, the gage would contain a ∅.501-in. D pin, which corresponds to the MMC size of the hole. This gage requires perfect perpendicularity when the hole is ∅.501 in. (at MMC) and is a more critical MMC callout.

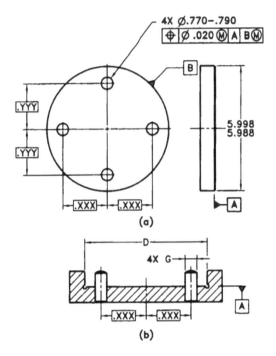

(a)

(b)

FIGURE 9-13   A less critical (MMC) outer diameter datum feature.

This ∅.501-in. D gage element would include the ∅.501-in. Go gage listed above in this case.

The use of perpendicularity tolerances on part datum features enables the product designer to determine the exact size or the fit allowance of the gage datum element and thus allows a variety of fits between gage datum elements and part datum features. MMC gages may "shake" or move on parts when part datum features are not at MMC and are therefore less critical than RFS gages.

Figure 9-13(a) illustrates a part that has MMC specified for the outer diameter of part datum feature B. Because no perpendicularity tolerance has been specified for part datum feature B from part datum feature A, the gage diameter D is 5.998 in., which is the MMC size of part diameter B [Figure 9-13(a)]. Gage diameter D would be 5.999 in. if part diameter B had a 0.001-in. perpendicularity tolerance at MMC specified in relation to part datum feature A. The four fixed gage pins in the gage shown in Figure 9-13(b) are ∅.750 in., derived as follows:

  ∅.770 in., at MMC

− ∅.020 in., positional tolerance specified for the holes

  ∅.750 in., fixed-gage-pin size G

For this operation, three gages are required: a ⌀5.988-in Not Go ring gage (note that the ⌀5.998-in. gage datum element is also the Go gage); a ⌀.770-in. Go plug gage; and a ⌀.790-in. Not Go plug gage.

9.4.3 Independent Hole Patterns

Figure 9-14(a) displays the entire periphery of a noncircular part datum feature modified with the MMC symbol. The gage [Figure 9-14(b)] will consist of four rails (a nest) 4.010 in. by 2.010 in., the MMC size of the part datum feature. Unless otherwise specified on drawings invoking the Y14.5M standard, all dis-

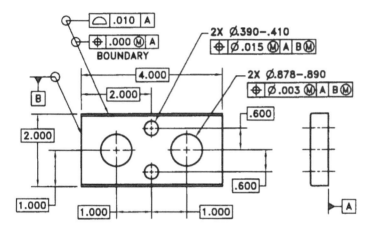

FIGURE **9-14(a)**   A less critical (MMC) noncircular datum feature (part detail).

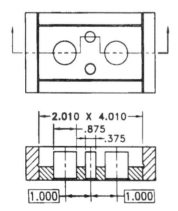

FIGURE **9-14(b)**   Gage detail for Figure 9-14(a).

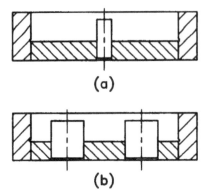

Figure 9-15 Functional gages for separate gaging of independent part features.

similar hole patterns on the same part must be gaged as one pattern if they are referenced to the same DRF with identical modifiers.

The drawing in Figure 9-14(a) may, however, specify with a note that the ∅.390–.410-in. hole patterns and the ∅.878–.890-in. hole patterns may be gaged separately. The note, specifying "separate requirements," would result in two gages: one for the ∅.390–.410-in. hole pattern [Figure 9-15(a)] and one for the ∅.878–.890-in. hole pattern as shown in Figure 9-15(b). This note would allow some independence of the hole patterns, and the two gages would not be as restrictive as the single gage depicted in Figure 9-14(b). It would be applicable if two separate items were assembled in the ∅.390–.410-in. and the ∅.878–.890-in. holes.

Plug and Snap Gages Required

1. ∅3.990-in. Not Go snap gage for minimum length of datum B
2. ∅1.990-in. Not Go snap gage for minimum width of datum B
   (Note: Maximum or Go dimensions of datum B are incorporated in the receiver gages in Figure 9-15.)
3. ∅.390-in. Go plug gage
4. ∅.410-in. Not Go plug gage
5. ∅.878-in. Go plug gage
6. ∅.890-in. Not Go plug gage

9.4.4 Two Critical Datum Features

Figure 9-16(a) shows two part datum features, diameter B and slot C, which control the radial and angular location of the part features. Because part datum

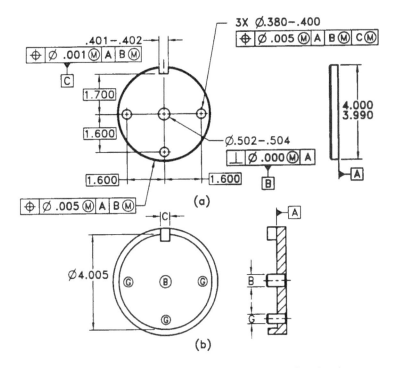

FIGURE 9-16 Two critical (MMC) datum features and functional gage.

features B and C are modified with MMC, the gage in Figure 9-16(b) will have a ∅.502-in. fixed-gage-pin B; and because the part slot, datum C, also has a perpendicularity of 0.001 in. to datum surface A, the size of gage element C will be 0.401 in. minus 0.001 or 0.400 in. The 0.001-in. perpendicularity tolerance is one way of specifying a "shake" gage datum element in that the gage element very likely will never fit too tightly in the part slot. The magnitude of the gage "shake" can be easily specified, as previously mentioned, by using various perpendicularity tolerances on part datum features. Thus, on the gage, diameter B equals 0.502 in., and width C equals 0.400 in. The three fixed pins G will be ∅.375 in., determined by subtracting the 0.005-in. positional tolerance at MMC from the MMC hole size of ∅.380 in.

Plug and Ring Gages Required

For datum C:

1. 0.401-in. Go
2. 0.402-in. Not Go

For datum B:

   3. ∅.504-in. Not Go

For the three clearance holes:

   4. ∅.380-in. Go
   5. ∅.400-in. Not Go

The outer diameter has a position tolerance of ∅.005 in. for part datum feature B at MMC. The inner diameter of the gage datum element that contacts this outer diameter will be

     ∅4.000 in., MMC size of the outer diameter

+ ∅ .005 in., positional tolerance

     ∅4.005 in., inner diameter of the gage

Two ring gages are required for the part's outer diameter: a ∅4.000-in. Go gage and a ∅3.990-in. Not Go gage.

The gage in Figure 9-16(b) properly combines several gaging operations on one gage. If the positional tolerance specified for the outer diameter, however, was only 0.002 in., and if this outer diameter approached 4.000 in. (its MMC size), and if part datum feature B were ∅.504 in., the outside ∅4.000 in. would assume the datum function and all holes would be located from this erroneous datum feature. The gage designer must watch for such pitfalls and use two separate gages when this problem can occur.

9.4.5 Multiple Datum Features, with Independent Hole Pattern

Figure 9-17(a) shows a hole pattern related to two part datum features, tabs B and C, at MMC.

Design Intent. The part will assemble into a mating configuration so designed that tabs B and C will be the key alignment features, with surface D the end locator. The six clearance holes will merely hold the part in place at assembly.

Functional Gage Design. Figure 9-17(b) shows the gage for the part in Figure 9-17(a). The slots in this gage are 0.4005 in. since part datum features B and C have a perpendicularity tolerance of 0.0005 in. at MMC, derived as follows:

     0.4000 in., MMC size of tabs B and C

+ 0.0005 in., perpendicularity at MMC specified for B and C

     0.4005 in., gage slot sizes B and C in Figure 9-17(b)

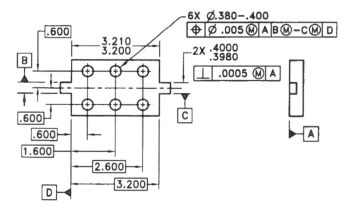

Figure 9-17(a) Two critical (MMC) datum features, with independent hole pattern, and functional gage (part detail).

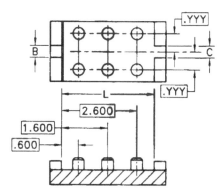

Figure 9-17(b) Gage detail for Figure 9-17(a).

The distance L between gage rails is 3.210 in., the MMC length of the part. Gage length L is not a gage datum element but merely a Go length gage. The diameter of the six fixed-gage pins is derived as follows:

  ∅.380 in., MMC size of holes

 − ∅.005 in., positional tolerance at MMC

  ∅.375-in. gage pins

The part must be placed in the gage against datum features A and D [Figure 9-17(a)]. This must be verified by suitable mechanical means.

Plug and Snap Gages Required

1. 0.4000-in. Go to check datums B and C
2. 0.3980-in. Not Go to check datums B and C
3. 3.200-in. Not Go [the 3.210-in. Go gage for dimension is incorporated in the receiver gage in Figure 9-17(b)]

The 0.0005-in. perpendicularity tolerance for datum tabs B and C is incorporated in the functional gage in Figure 9-17(b).

9.4.6 Datum Features Related to Primary Datum Plane

Figure 9-18(a) depicts a part that contains part datums B and C and primary datum plane A. This method of dimensioning parts from two holes is quite common because it is practical to locate parts on tools that contain dowel pins. In this example pin B is smaller than pin C, to force datum precedence.

Figure 9-18(b) shows the multipin gage that meets the drawing requirements. The gage pins are determined as follows:

1. B gage datum element

    $\varnothing$.3755 in., at MMC

    $- \varnothing$.0005 in., perpendicularity tolerance at MMC

    $\varnothing$.3750 in., at B

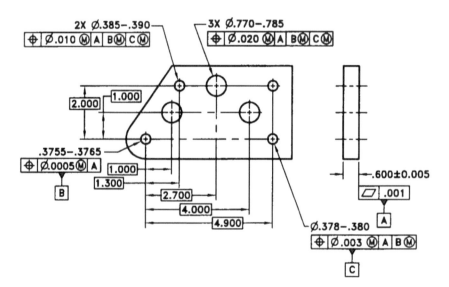

FIGURE 9-18(a) Part with two holes as datum features related to primary datum.

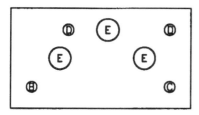

Figure 9-18(b)   Multipin functional gage for 9-18(a).

2. C gage datum element

    ∅.378 in., at MMC

    − ∅.003 in., positional tolerance at MMC

    ∅.375 in., at C

3. D gage elements

    ∅.385 in., at MMC

    − ∅.010 in., positional tolerance at MMC

    ∅.375 in., at D

4. E gage elements

    ∅.770 in., at MMC

    − ∅.020 in., positional tolerance at MMC

    ∅.750 in., at E

Plug Gages Required

1. ∅.3755-in. Go for datum feature B
2. ∅.3765-in. Not Go for datum feature B
3. ∅.378-in. Go for datum feature C
4. ∅.380-in. Not Go for datum feature C
5. ∅.385-in. Go for ∅.385−.390-in. clearance holes

6. ∅.390-in. Not Go for ∅.385–.390-in. clearance holes
7. ∅.770-in. Go for ∅.770–.785-in. clearance holes
8. ∅.785-in. Not Go for ∅.770–.785-in. clearance holes

9.4.7 Three-Hole Pattern and External Datum Feature

Figure 9-19(a) shows a radial three-hole pattern dimensioned from primary part datum feature A and diameter B at MMC.

Design Intent. The part must fit into a sleeve against surface A and be pinned in location through three holes. Part datum feature B must be perpendicular to part datum feature A at MMC.

Functional Gage Design. Several dial indicators could be mastered so that they would all read zero when the part is fully inserted into the gage [Figure 9-19(b)]. The 2.000-in. internal diameter in the gage is a Go gage that checks both the maximum size of the part's outer diameter and its perpendicularity to datum surface A. The three G gage elements are ∅.313 in., determined as follows:

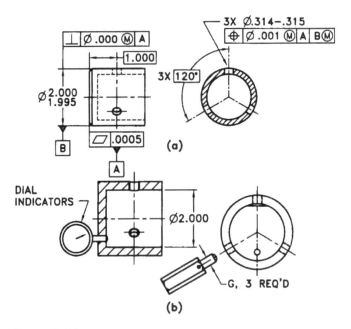

FIGURE 9-19 Part with radial three-hole pattern and external datum feature and functional gage.

Ø.314 in., clearance hole at MMC

− Ø.001 in., positional tolerance specified at MMC

Ø.313 in., G gage basic

Plug and Ring Gages Required

1. Ø.314-in. Go
2. Ø.315-in. Not Go
3. Ø1.995-in. Not Go snap gage

The same gage design technique is used regardless of the number of radial holes in the pattern. The same basic gage would also be used if the three holes shown in Figure 9-19(a) were inclined at some basic angle, other than 90°, to part datum feature A, except that the gage bushings would be placed at this same angle.

9.4.8 Three-Hole Pattern and Internal Datum Feature

Figure 9-20(a) portrays a radial three-hole pattern dimensioned from part datum feature A and inner diameter B at MMC.

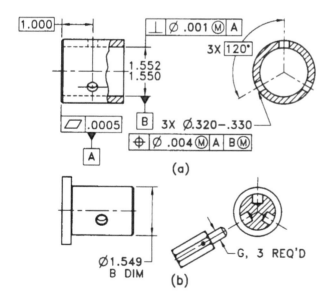

FIGURE 9-20 Part with radial holes and an internal datum feature and functional gage.

Design Intent. The part must fit over a plug against surface A and be fastened with three bolts. Part datum feature diameter B need not be perfectly square at MMC, because it is allowed 0.001-in. perpendicularity at MMC to part datum feature A.

Functional Gage Design. The gage [Figure 9-20(b)] is a shouldered plug with an outer diameter of 1.549 in., determined as follows:

     Ø1.550 in., the MMC size of the part datum feature

   − Ø <u>.001</u> in., perpendicularity tolerance specified at MMC

     Ø1.549 in., gage element B

For acceptance, part datum feature A must fully contact gage datum element A. Suitable mechanical verification is required. The G gages are Ø.316 in. The same basic gage would also be used if the three holes shown in Figure 9-20(a) were inclined at some basic angle other than 90° to part datum feature A, except that the gage bushings would be placed at this same angle.

Plug Gages Required

1.   Ø1.550-in. Go
2.   Ø1.552-in. Not Go
3.   Ø.320-in. Go
4.   Ø.330-in. Not Go

9.4.9   Cylindrical Part with Two-Pin Patterns

Figure 9-21(a) shows a cylindrical part with one two-pin pattern and one four-hole pattern. The part datum for these patterns is the axis established by the 1.000- to 1.005-in. inner diameter at MMC. The two end surfaces have not been designated as part datum features and will only be contacted at one high point when the gage [Figure 9-21(b)] is placed on the part.

Design Intent. The mating part passes through part datum feature diameter A [Figure 9-21(a)]. The two patterns must be located from the inner diameter of part datum feature A when this feature is at MMC (1.000 in.), and the part datum feature must be straight at MMC (when it is finished to 1.000 in.). The drawing note states that the two-pin and four-hole patterns are not radially related (they do not have to align for the parts to function at assembly).

Functional Gage Design. The gage datum element diameter (1.000 in.) is the MMC size of part datum A. Two 0.254-in. internal diameter bushings are placed on gage ring 1, which fits closely over the 1.000-in. receiver diameter. Four Ø.250-in. gages fit through the four close-fitting bushings on the gage ring

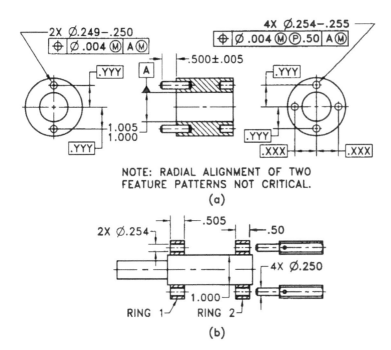

NOTE: RADIAL ALIGNMENT OF TWO
FEATURE PATTERNS NOT CRITICAL.

(a)

(b)

FIGURE 9-21 Part with two feature patterns and functional gages.

2 element, which also fits closely over the 1.000-in. receiver gage datum element diameter. The two rings, which control the two- and four-hole patterns on the part, are free to rotate on the 1.000-in. receiver diameter since pattern alignment is not critical. Ring 1 is at least 0.505 in. thick, which is the maximum height of the two Ø.249–.250-in. pins, and ring 2 is at least 0.50 in. thick, the maximum thickness of the mating part. If the drawing did not state that the two-feature patterns could rotate (or misalign), the two rings should be aligned or oriented with keyways or pins.

Plug and Ring Gages Required

1. Ø.250-in. Go ring
2. Ø.249-in. Not Go ring
3. Ø1.005-in. Not Go plug (The Ø1.000-in. receiver gage element checks the Ø1.000-in. "Go" requirement.)
4. Ø.254-in. Go plug
5. Ø.255-in. Not Go plug

9.4.10   Two Radial Patterns of Pins and Slots

Figure 9-22(a) shows two radial patterns of features (pins and slots).

Design Intent.   The radial and angular orientation of these part features is critical in relation to part datum feature B, but their location from part datum feature A is not critical. The outer diameter of the part must be positioned within ⌀.005 in. at MMC to part datum feature B at MMC.

Functional Gage Design.   The gage [Figure 9-22(b)] is one unit since the pins and slots are related and are actually one pattern of features dimensioned from part datum B at MMC. This is the result of using identical DRFs and modifiers that create simultaneous requirements.

1.   The gage datum element B is ⌀1.9995 in. because the part datum feature has a perpendicularity tolerance of ⌀.0005 in. in relation to part datum feature A at MMC.

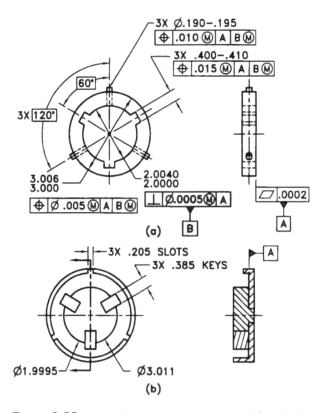

FIGURE 9-22   Part with radial pins and slots and functional gage.

⌀2.0000 in., part datum feature B at MMC

− ⌀ .0005 in., perpendicularity tolerance at MMC

⌀1.9995 in., gage datum element

2. Pin location (radial) is checked with three 0.205-in.-wide slots.

0.195 in., MMC of pin

+ 0.010 in., positional tolerance at MMC

0.205 in., gage element slot

3. Slot locations are checked with three 0.385-in. contacts.

0.400 in., MMC of slots

− 0.015 in., positional tolerance of slots

0.385 in., gage key element

4. Part outer diameter is checked for position with the ⌀3.011-in. gage inner diameter, determined as follows:

⌀3.006 in., MMC of part outer diameter

+ ⌀ .005 in., position allowed at MMC

⌀3.011 in., gage element

Plug Gages Required

1. ⌀.195-in. Go
2. ⌀.190-in. Not Go
3. 0.400-in. Go
4. 0.410-in. Not Go
5. ⌀2.0000-in. Go (required because functional gage is 1.9995 in.)
6. ⌀2.0040-in. Not Go
7. ⌀3.006-in. Go (required because functional gage is 3.011 in.)
8. ⌀3.000-in. Not Go

9.5 REVIEW OF PRINCIPLES AND APPLICATIONS

The material in this chapter helps the product development team determine tool and gaging requirements; it allows the gage designers to determine the form, relationship, and location of gage elements; and it shows product designers how to dimension and tolerance engineering drawings so that practical tools and gages may be used.

A set of generalized rules to aid in gage design can be gleamed from the examples:

1. The specified positional tolerance should be added to the MMC size of male part features to obtain the basic female size of the functional gage element.
2. The specified positional tolerance should be subtracted from the MMC size of female part features to obtain the basic male size of the functional gage element.
3. Part datum features modified with RFS must be "centered" in the gage.
4. Part datum features modified with MMC must be contacted by gage datum elements that are the MMC size of these part datum features.
5. Part datum features modified with MMC that have perpendicularity tolerances also modified with MMC are contacted by gage datum elements that never freeze on the part datum feature. Such a gage could always shake on the part.

10

Functional Gage Tolerancing

10.1 INTRODUCTION

Chapter 9 presents fundamentals of functional gage design implicitly assuming a tolerancing scheme. While the tolerancing rules may give a superficial appearance of being simple and easy to apply, some hidden concerns should be explicitly addressed. The driving force behind interest in these techniques is gaging policy. The original discussion starts in Chapter 5, where it is argued that a specific gaging policy must be selected and included as part of contractual obligations. There are also related discussions associated with measurement uncertainty and decision rules for determining conformance.

The following material illustrates methods that may be used to apply tolerances to functional gages. The examples suggest issues that must be pursued when real gages are to be built. At their root these issues involve the mechanics of implementing gaging policy and the resulting impact on acceptance rates. The risk associated with the final choice of policy can be reduced by an understanding of the implications of the different tolerancing techniques. Ultimately the decisions are economic in nature.

10.2 GAGING ELEMENT SIZE AND MATERIAL MODIFIERS

Any physical gage requires selection of gaging element size, tolerances of size and geometry, and appropriate material modifiers. In an ideal world, the gage design would be derived from workpiece specifications using simple rules-of-thumb and quickly executed procedures. Unfortunately, simple rules inevitably have hidden consequences that should be considered when selecting the gaging policy.

A variety of gaging policies may be implemented when designing functional gages. The default policy in the United States might be considered to be absolute (pessimistic) tolerancing, as described in Sec. 5.6. This policy is intended to reject all part features not within specification.

There are costs, both apparent and hidden, associated with any tolerancing scheme. In the case of absolute tolerancing, a small percentage of borderline part features technically within tolerance may be rejected. This cost is considered to be acceptable in order to fulfill the policy's intent, to accept only good parts.

The following analysis is restricted to absolute tolerancing procedures and their effect on production yield. Similar analyses could be performed for the other tolerancing policy alternatives but are excluded for brevity. Included examples illustrate the range of results that occur when using the three material condition modifiers allowed by ASME Y14.5M. Each design choice results in a different conformance zone. The examples are used to verify whether a particular material condition choice in conjunction with absolute tolerancing succeeds in rejecting all out-of-tolerance parts (i.e., conforms to the gaging policy).

The design criteria driving the examples utilize the virtual condition boundary of the workpiece feature to establish one of the gaging element's boundaries—the inner locus boundary. The locus boundary concept is used to illustrate each example and discussed in some detail with an alternate approach to the analysis in Sec. 10.6. This gage boundary is the starting point in determining the acceptance region. The gage boundary is then used to calculate the gaging element size and appropriate tolerances based on the selected material condition modifier.

The functional gage examples are constructed for a prismatic part with four ∅.272-in. holes controlled with a positional tolerance using the maximum material condition (MMC) modifier (Figure 10-1). At one level the analysis could be based on statistical approaches that more closely model the actual variation that would be encountered. Because these are three-dimensional examples, the underlying variation would probably follow a multivariate statistical distribution. However, this introduces statistical and mathematical complications that are not easily resolved. It should suffice to say that such analysis is well beyond the intent of this book. Rather, the analysis model assumes that the positional errors of the

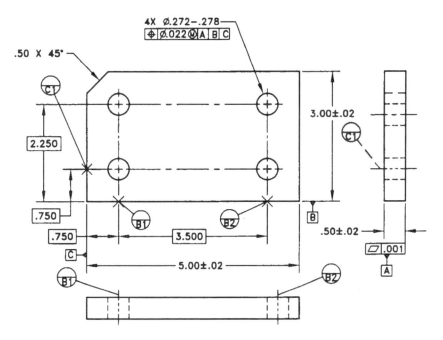

FIGURE 10-1 Workpiece example for gage analysis.

two mating features, the gage pin and the mating hole, are at their respective extremes (i.e., combinations of size and position). These extremes are then used to determine the range of part geometry the gage accepts.

There is little likelihood that the specific combination of positional errors illustrated in the examples will occur. These limiting values require not only that the extreme size and positional error values occur simultaneously on mating parts, but also that the positional errors of both features occur along the same line and with the same directional sense. While this particular combination of errors would seem to be of little practical consequence in view of its small probability of occurrence, the selected tolerancing policy (i.e., the absolute) purports to accept good parts and reject bad parts. With this in mind, it is appropriate to explore the limiting cases to see if the design rules meet the intent of the gaging policy and understand their limitations if the policy is not met. The goal is to be able to make informed choices when designing and building the gage.

10.3 WORKPIECE EXAMPLE

As mentioned, the workpiece used for the analysis is a prismatic part shown in Figure 10-1. The four holes drilled in the part are ∅.272 in., with a size tolerance of +0.006 and −0.000. This size was chosen to take advantage of commercially available tooling (letter size drill H) and uses tolerance estimates for drilled holes cited by a number of sources (Truks, 1987). The four holes are completely located by basic dimensions within a datum reference frame.

The location controls are based on a floating fastener calculation using a ¼-20 UNC fastener, with an MMC size of ⌀.250. The floating fastener calculation yields

⌀.272, MMC hole

−⌀.250, MMC fastener

⌀.022, positional tolerance

The complete callout for the four holes is

$$4X\ \varnothing.272\ {}^{+.006}_{-.000}$$

| ⌀ | ⌀.022 Ⓜ | A | B | C |

The size and positional callouts yield a part feature virtual condition equal to

virtual condition = MMC size of the hole

− geometric tolerance at MMC

virtual condition = ⌀.272 − ⌀.022 = ⌀.250

The other boundary of concern is called the *resultant condition*. For this specific application, the resultant condition is calculated as follows:

resultant condition = LMC + geometric tolerance at MMC

+ difference in size between MMC and LMC

= ⌀.278 + ⌀.022 + ⌀.006

= ⌀.306

Both the virtual condition and resultant condition boundaries are used to evaluate gage effectiveness in the design cases analyzed ahead and in the appendix to this chapter.

All the examples assume the part features have been checked to determine if they are within the size specification. Only components that pass this first level of control are presented to the functional gage. The generic form of the functional gage looks like the illustration in Figure 10-2. Drawing details reflecting tolerance choice and modifier would be added to describe the completed gage after a dimensioning and tolerancing scheme is selected. At this point the figure is intended only to provide the general configuration of the functional gage that underlies each of the following examples.

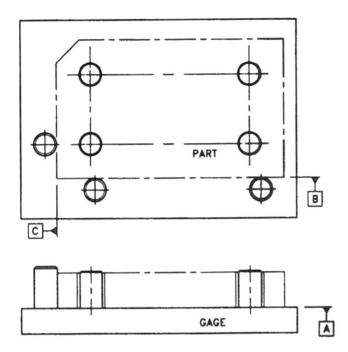

FIGURE 10-2   General form of gage used to inspect features of workplace.

The analyses in the next section and in the appendix (Secs. A10.1 through A10.4) show how to determine the theoretical conformance range created by a particular gage example. With the exception of size and position tolerance on the gage pin, this part of the chapter does not provide detailed comments on gage manufacturing issues. Sections 10.7 and 10.8 comment on manufacturing concerns that must be considered if a physical gage is to be fabricated, some having a direct impact on the design's effectiveness.

10.4   ZERO POSITIONAL TOLERANCE AT LMC

The first example uses the LMC modifier with a zero positional tolerance for the specification on the gage pin. The perfect gage pin is $\varnothing.2500$ when located at true position and at a basic 90° angle to datum A. This virtual condition size is the starting point to construct a physical gage element.

If a zero positional tolerance at LMC is applied to the gaging element, the total tolerance available to the gage pin—under one set of rules—becomes a combination of size and positional tolerance. The part tolerance is

0.0060, the part's size tolerance

+0.0220, the part's positional tolerance at MMC

0.0280, total available part feature tolerance

As the gage element departs from LMC, the departure in size can be captured and used to accommodate position error. As a result, there is a one-to-one exchange between size departure from LMC and the available positional error allowed on the gage pin.

To calculate the tolerance on the gage pin, the ⌀.028 total available tolerance on the part feature is used as the starting point to allocate a portion of total product tolerance to the gaging element. Using the traditional gagemaker's rule of 10%, the tolerance allocated to the gage pin is

total feature tolerance × 0.10 = total gage element tolerance

For this particular example,

total tolerance on gage pin (size and position)

$$= 0.028 \times 0.10 = 0.0028$$

The full ⌀.0028-in. tolerance will be shown as the size tolerance on the gage pin. The complete callout for the gage pin is

$$4X \ \varnothing.2500 \ {{+.0028} \atop {-.0000}}$$

| ⌀ | ⌀.0000Ⓛ A | B | C |

We now explore the critical combinations of pin size and location that can be realized in building the gage. In all subsequent figures, the reference for the analysis is the true position of the part feature as indicated by "TP" in the diagrams.

10.4.1 LMC Gage Pin at True Position

This is an LMC pin located at true position. The pin size is ⌀.2500, the virtual condition of the part features, and the gage pin boundary imposed by our design rule. In theory, this will accept parts at their virtual condition limit as established by specification.

10.4.2 MMC Gage Pin at True Position

This is a ⌀.2528 pin that is larger than the virtual condition of the part feature being inspected and is located at true position. If the gage pin is fabricated at the MMC size, the gage pin will be ⌀.0028 larger than the virtual condition size of the part feature established by the designer. In this situation, the gage will reject some parts that are actually within the acceptable range of design values—

good parts are being rejected, those with generated boundaries falling between the part's virtual condition size (∅.2500) and the MMC gage pin at true position (∅.2528).

This range of values is controlled by the gage pin tolerance. At the relatively large pin tolerance—in gagemaker's terms—shown in the example, the tolerance on the pin could be reduced, still stay within the normal range of commercial components, and remain faithful to established gagemaker tolerance classes. At this nominal size (∅.250 in.), gagemaker tolerances run from 0.000020 in. (class XX) to 0.000100 in. (class Z). In this example, the gage designer has significant discretion to reduce the range of values within which good parts might be rejected by reducing the size tolerance of the pin.

10.4.3 MMC Gage Pin with Maximum Positional Error

With a gage pin fabricated at its MMC size, this case looks at a pin located at the maximum displacement from true position, ∅.0028, allowed by the gage design. This shift is illustrated in Figure 10-3 by displacing the gage pin center 0.0014 to the left. The figure illustrates the situation using radial values. With the exception of Secs. 10.6 and A10.4, all subsequent tolerance analyses use radial values for descriptive purposes.

Boundary of MMC pin (∅.2528)	0.1264
Shift of pin center allowed by maximum positional error	−0.0014
Distance from true positional to pin boundary	0.1250
Radius of LMC hole (∅.278)	0.1390
Distance from true position to pin boundary	−0.1250
Difference between true position and hole center	0.0140

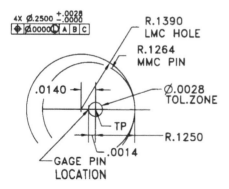

FIGURE 10-3   Functional gage with zero at LMC positional tolerance.

TABLE 10-1   Gage Pin Sizes for Various Design Methods

Gage design method	LMC gage pin at true position	MMC gage pin at true position
Hole virtual condition	0.2500	0.2500
LMC—zero tolerance	0.2500	0.2528
RFS	0.2511	0.2522
MMC—10% rule	0.2500	0.2506
MMC—zero tolerance	0.2500	0.2528

The result, 0.0140, is the radial shift allowed to the center of an LMC hole on the part. This shift is with respect to the true position location and can be converted to the diametral tolerance zone for comparison with the part specification. The resulting diametral zone is

$$\text{positional tolerance zone} = 0.0140 \times 2 = \varnothing.0280$$

This matches the $\varnothing.028$ maximum positional tolerance allowed to the hole when it is fabricated at its LMC size.

10.5 RESULTS

The results of the preceding example and the remaining design alternatives explored in the appendix are summarized in Tables 10-1 and 10-2 and Figure 10-4. For purposes of discussion, gage effectiveness is defined as the proportion of the product specification that forms the conformance range established by the gage design. A gage design that allows use of a greater part of the specification range is more effective than a similar design using less of the specification range.

In theory, the LMC gage specification with a zero positional tolerance offers an effective design technique to implement the absolute gaging policy. Such a design accepts the virtual condition part if the gage pin is fabricated at its LMC size and located at true position. However, this requires a perfectly positioned LMC pin to become reality. Because such perfection is highly unlikely, manufac-

TABLE 10-2   Maximum Positional Error for Gage Pin at LMC and LMC Hole

Part specification	0.028
LMC Gage—zero tolerance	0.028
RFS Gage	0.028
MMC Gage—10% rule	0.0308
MMC Gage—zero tolerance	0.0308

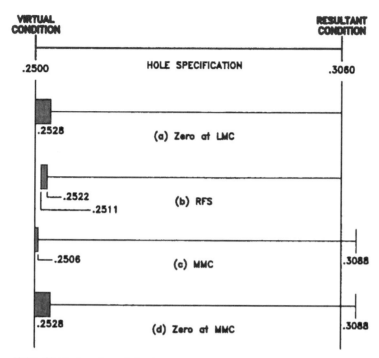

Note: Shaded regions indicate area
of product tolerance affected by
gage pin size.

FIGURE 10-4   Results of tolerance analysis.

turing has to select the gage pin toward MMC size to make a nonzero positional tolerance value available for fabrication. At the opposite end of the specification range, the gage will not accept parts beyond the maximum positional tolerance of ∅.028 allowed by specification.

The RFS gage (Sec. A10.1) provides a somewhat less effective technique to achieve the same results as the LMC gage. The RFS gage will encroach into the part tolerance—the lower specification limit—in order to accept good parts and reject bad ones. To accomplish the latter, some good parts are rejected at the inner locus boundary (i.e., a true-position LMC gage element of ∅.2511 versus a part virtual condition of ∅.2500).

The MMC gage (Secs. A10.2 and A10.3)—using either 10% of the part's positional tolerance or a zero value for a gage tolerance—has similar characteristics as the RFS gage at the lower specification limit. These gage designs encroach on the part to an extent dictated by the finished size of the gage element. The designs match the virtual condition of the workpiece but allow acceptance of extreme parts (i.e., LMC holes at maximum positional error) outside specification.

10.6   ALTERNATE FORM OF ANALYSIS

Radial values are used in the examples to provide better graphical demonstration of the results. This is probably not the way it would be done by GD&T practitioners. The more likely analysis would use diametral calculations consistent with the typical cylindrical shape of the positional tolerance zone. To illustrate the more common approach, the example in Sec. 10.4 is repeated using more conventional methods. A similar example is found in Sec. A10.4. The calculations require determining an inner boundary of the gage pin and the outer boundary of the inspected hole. The resulting boundaries are more correctly referred to as locus boundaries. The term "locus" is used in its mathematical form as "the curve or other figure composed of all the points which . . . are generated by a point, line, or surface moving in accordance with defined conditions" (Oxford, 1993). The boundaries are related to the physical feature but are not limited to the obvious boundary that circumscribes part material. Detailed examples of boundary calculations for a variety of situations can be found in Meadows (1998).

The analysis requires calculation of the boundaries as follows:

gage pin inner locus boundary

$$= \text{LMC} - \text{geometric tolerance specified at LMC}$$

For this example, the specified value of the geometric tolerance is zero at LMC size:

gage pin inner locus boundary = $\varnothing.2500 - \varnothing.0000 = \varnothing.2500$

The analysis also requires calculation of the outer locus boundary of the hole. Referring to the part specification, the size of the feature at LMC was $\varnothing.278$, the geometric tolerance was $\varnothing.022$, and the available bonus was $\varnothing.006$.

hole outer locus boundary = $\varnothing.278 + \varnothing.022 + \varnothing.006 = \varnothing.306$

This outer boundary value represents the boundary for a hole produced at maximum positional error allowed by specification. The intent of the following analysis is to determine if either of the two gaging cases meets this specification or deviates from the desired results.

The difference between the inner locus boundary of the pin and the outer locus boundary of the hole is used to establish the theoretical acceptance region of the gage. The tolerance range of workpiece accepted by the gage is illustrated

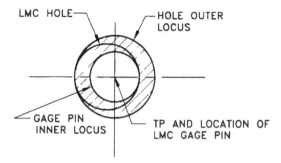

FIGURE 10-5 Functional gage analysis using diametral specifications.

in Figure 10-5. It shows the acceptance region created by these two boundaries and within which the inspected feature must lie. The actual displacement of the physical feature relative to the center of the gaging area (i.e., true position of the hole) depends on the combination of size and positional tolerance at which the gage pin and the hole are produced. The maximum displacement for this gage example occurs for the combination of a part produced at the LMC hole (∅.278) and an LMC gage pin with maximum allowed displacement from true position. The latter feature generates the gage pin's inner boundary.

Figure 10-5 takes the radial analysis methods and uses the equivalent diametral values. In the figure the inner boundary for the gage pin is located at the true position for the gage, which is also the true position of the hole. The intermediate-sized circle represents the LMC hole displaced the maximum distance allowed by the gage pin. The largest circle is generated when the LMC circle is rolled around the inner boundary gage pin. This outer circle is the locus of points described by the unusual combination of an LMC hole mating with an LMC gage pin, each with maximum positional error and shifted in direction and sense to create the worst-case situation. The combination yields

maximum positional tolerance

= [2(LMC hole − ½ gage pin's inner boundary)] − LMC hole

A little algebra shows that this reduces to an equation that yields the same results as the radial analyses:

maximum positional tolerance

= LMC hole − gage pin's inner boundary

For a gage with zero tolerance specified at LMC,

maximum positional tolerance = ∅.278 − ∅.250 = ∅.028

As anticipated, this numerical value matches the maximum positional tolerance allowed if the part feature is produced at LMC. Figure 10-5 illustrates this case and shows the acceptance region created by the design.

It should be mentioned that when this technique is applied to gage design using any feasible alternative specification, all critical combinations of size and tolerance should be investigated. It is the designer's responsibility to ensure that the consequences of the design are fully understood and that all parties reach a consensus on acceptability of the design.

10.7 FITS AND ALLOWANCES

While all of the examples have assumed a worst-case analysis with the parts mating at their respective virtual condition sizes, this technique may not be acceptable in all situations. Many products utilize specified fits and allowances to achieve the desired function. If the gage analysis involves a product that incorporates predetermined allowances, then the gage calculations must be altered.

Techniques can be developed to accommodate this situation. One approach that could be used involves allocating a portion of the allowance to each member of the mating pair. One of the authors helped a large manufacturer explore this type of technique. After extended discussion, the technical team decided to apply 40% of the allowance to each of the mating parts and retain 20% of the allowance as a buffer to ensure proper assembly.

10.8 BUILDING THE GAGE

At some point either a gage or a fixture will be built. This section looks at the ramifications in moving the gage design from concept to metal. The discussion is framed around our sample part (Figure 10-1) with a specific gage design used to illustrate tolerance considerations and their effect on gage fabrication.

10.8.1 Machine Tool Capabilities

In our example, fairly large tolerances have been specified. These values are consistent with the philosophy of providing justifiable tolerances that are as large as possible within functional requirements, providing manufacturing with as much latitude as feasible. The $\emptyset.022$-in. positional tolerance chosen for Figure 10-1 falls into the ISO fine tolerance category for clearance holes (floating-fastener case). Since most current machine tools are capable of holding tolerances to ± 0.00011 in., they can locate bored holes for gage pins or bushings to positional tolerance diameters of 0.0003 in., far closer than needed to fabricate the gages illustrated in Figures 10-6 through 10-9. In addition, commercially available radial relief micro boring bars provide more than sufficient capability for the tolerances specified in the workpiece and gage illustrations. These boring bars are designed for CNC porting applications that range in diameter from 0.015 to 0.490 in. The boring bars contain tenth-adjusting boring heads capable of moving the tool 0.00005 in. radially.

With this available technology, these gage designs may never come close to rejecting parts under any of the material conditions, because production machine tools have improved by at least an order of magnitude in the past 20 years. In the case of the example gages, the 10% rule may not make sense, as the clearance holes and gage pins may never make contact. As a consequence, little gage wear will likely occur.

Tolerances applied to the datum feature simulators on the gage are of paramount interest to the gage designer because of their effect in creating an apparent increase in part tolerance. These apparent increases in part tolerance may result as the gage departs from the perfect geometry by

Tolerances of form placed on gage surfaces
Tolerances of size, form, orientation, and location applied to features of size that comprise the fixture and gaging elements
Tolerances associated with relationships that create the DRF

When a part is assembled over the four ∅.2500-in. gage pins during inspection, the part datum targets must contact the three ∅.2000-in. dowel pins on the gage. The positional tolerances on these target pins may shift the part left to right if the tertiary datum pin is not the nominal size or at the basic location. The part may also translate up or down or assume an angular orientation if the two pins creating the secondary datum are not in line or the same size. The designer and the gagemaker should be alert to this situation. While this may appear as stating the obvious, it is a possible cause of part rejection if the three ∅.2000 dowel pins on the gage are not functional (i.e., not representative of mating part geometry). This functional consideration has been emphasized throughout the book, and it is expected that the designers are now well versed in the concept that gages should represent mating parts.

It is pointed out earlier in this chapter that there is a very small probability of occurrence associated with the worst-case situations used in the analyses. The basis for this conclusion is the random nature of the errors being considered. One caveat that should perhaps be mentioned is that once the gage is fabricated, the errors associated with a particular physical gage are no longer random in nature. Randomness is still present in the workpiece/gage system during the inspection process, but this effect is primarily due to the random distribution of errors associated with part manufacture, not due to random effects present in the physical gage. As a consequence, acceptance probabilities are affected by the state of the gage immediately following fabrication and during its useful life—hence the need for formal maintenance and calibration procedures to ensure adequate control of the gage or fixture.

The bilateral tolerancing system, which in effect dictates RFS on everything associated with the gage assembly, still appears to be the norm. At least one automotive company utilizes this method coupled with notes such as "10% not to exceed ±0.12 mm" to place bounds on the available gage tolerances. The authors have found these RFS gage applications throughout industry with only one exception. This was at a government facility with extremely critical functional specifications imposed on manufacturing. The facility used LMC in a few applications but then discontinued the practice because the gage designers insisted that manufacturing worked only to bilateral tolerances at RFS. The same attitude was found in the automotive firm and makes it difficult to enforce GD&T on the tool drawings if these specifications are to be ignored.

Most gages are subsequently calibrated to bilateral tolerances (RFS). To the authors' knowledge, virtual condition has not been applied to gage feature calibration. It is possible that either actual local size or actual mating size is all that has been used in determining when a gage has worn out.

10.8.2 Single-Setup Gage Feature Manufacture

The control level that can be built into the finished gage is strongly influenced by the manufacturing process sequence. To illustrate this, two types of gages have been considered, fixed pin (case I) and removable pin (case II). In both design alternatives, the manufacture of these gages has been approached using the single-setup methods recommended by the authors. The accompanying design for the gage base shows a process DRF used to establish the inspection DRF. The former is composed of datum A, datum Y, and datum Z (see Figure 10-6). The inspection DRF is composed of datum A, datum B, and datum C, which are identified on the assembly drawings (Figures 10-7 and 10-8). The gage pins have been given tolerances to the minus side so that there will be no tolerance stackup.

The following operations provide gages for the workpiece shown in Figure 10-1.

Case I—Fixed-Pin Gage

Operation 1: Rough machine rectangular gage base to fit part in Figure 10-1.

Operation 2: Grind primary datum A to 0.0002 flatness. See Figure 10-6.

Operation 3: Clamp gage base on primary datum A, align, and machine surfaces Y and Z.

Operation 4: Zero machine tool off datums Y and Z for single-setup location of seven bores.

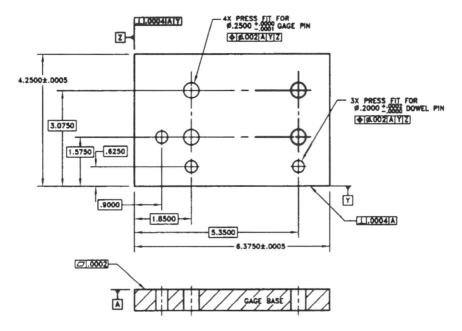

FIGURE 10-6   Gage base for single-setup manufacturing.

Operation 5: Drill and micro bore gage features for press fit of datum targets and gage pins.

Case II—Removable-Pin Gage   Repeat operations 1 through 5, but drill and micro bore the four gage features for slip renewable bushings. These operations will suit an RFS, MMC, or LMC gage, but the gage pins will be sized for a clearance fit into the set of four commercial bushings.

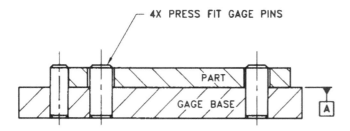

FIGURE 10-7   Gage and part assembly for case I.

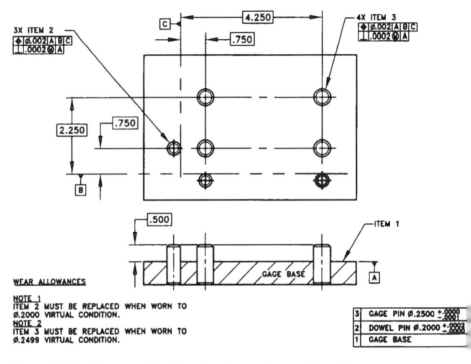

FIGURE 10-8 Gage assembly with calibration requirements (Case I).

10.8.3 Gage Assembly Operations

Case I—Fixed-Pin Gage

Operation 1: Place a set of ∅.2000- and ∅.2500-diameter gage pins in a freezer to soak.

Operation 2: Fit dowels and gage pins and press them into bores (Figure 10-7).

Operation 3: Check that dowel and gage pins are secure after soaking gage at 68°F.

Operation 4: Measure all pin locations and perpendicularity and record virtual condition.

Operation 5: Remove all pins that approach wear allowance.

Operation 6: Determine condition of wear allowance on each gage feature prior to use.

Operation 7: Determine calibration schedule on gages (Figure 10-8).

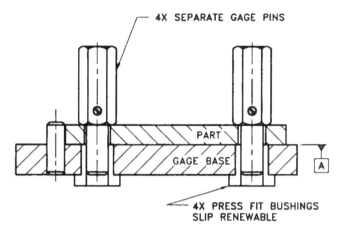

4X SEPARATE GAGE PINS

PART

GAGE BASE

A

4X PRESS FIT BUSHINGS
SLIP RENEWABLE

FIGURE 10-9   Bushed gage and part assembly for Case II.

Case II—Removable-Pin Gage   Repeat assembly operations 1 through 3 using bushings. Each bushing should be measured in operations 4 through 7 with the gage pin inserted into the bushing and the gage base. This should be done with the gage base and pins at 68°F. Then proceed with operations 4 through 7. Finish by fitting handles on the four gage pins as shown in Figure 10-9.

The same operator on the same machine tool should fabricate all the gage bases. The positional tolerances (Figures 10-6 and 10-8) can be easily held with current machine tools that guarantee ±0.00011 in. Note that several backup gages should be provided.

10.9   SUMMARY

The results of the four examples are shown in Table 10-2 and Figure 10-4. In all of these examples, relatively large tolerances are shown for the gage. It is recognized that these gage tolerances may be reduced significantly and still remain in the realm of commercially available gaging components or standardized gagemaker tolerances. However the effects are still the same.

All the example gage design analyses are based on worst-case situations whose probability of occurrence is small. However, it is worth noting that once the gage is built, a unique combination of size, location, and direction of error is built into the gage pin and becomes fixed rather than random. As a consequence, the random effects of mating the hole with the gage are now generated only by the holes presented to the gage and wear characteristics of a specific physical gage.

With the hole's virtual condition boundary used as the inner gage boundary, there is a gray zone at the lower specification limit where good parts may be rejected based on the actual gage pin size. The extent of this area is determined by the size and tolerance the gage pin is allowed as it is produced.

The LMC gage theoretically stays within the specification limits but requires an MMC size pin with the maximum positional error to accept parts made at the maximum positional tolerance for the hole (the hole resultant condition). Additionally, material must be added to the LMC pin to have any tolerance with which to make the gage. The result is that the conformance region is moved away from the lower specification limit.

The RFS gage protects the gaging policy by creating a conformance zone whose boundaries are contained inside the specification at the lower limit. The gage rejects technically good parts at the lower end of the specification zone to ensure that no bad parts are accepted.

The MMC gage with a nonzero positional tolerance meets the virtual condition boundary of the part with the gage pin at LMC. At the other end of the specification range, the MMC gage can accept parts larger than the hole's resultant condition—those parts with greater positional errors than allowed by specification. The extent of this situation depends on the size of the gage pin, the error associated with pin location, and the geometry of the hole presented to the gage.

An MMC gage with zero positional tolerance has a larger range where it may reject good parts when the gage pin is located at true position. This zone is created between the LMC size and the MMC size pin. At the other end of the specification range, this design technique has the same results as the MMC gage with a specified nonzero positional tolerance; it accepts parts with greater positional tolerances than allowed by specification.

There are many possible ways to calculate and partition tolerances. Each method will require analysis and documentation.

REFERENCES

Meadows, J. D., Geometric Dimensioning and Tolerancing, New York: Marcel Dekker, 1998.

Oxford University Press, The New Shorter Oxford English Dictionary, New York: Oxford University Press, 1993.

Truks, H. E., Designing for Economical Production, 2nd ed., Dearborn, MI: Society of Manufacturing Engineers, 1987.

Appendix 10.A

To avoid bogging down the reader in details, the remaining analyses used in the chapter and necessary to the comparisons documented in Tables 10-1 and 10-2 and Figure 10-4 have been placed in this appendix. The LMC example in Sec. 10.4 is joined with the following alternatives to show the range of options available to the designer.

A10.1   RFS CALLOUT ON GAGE PIN

The part virtual condition is again used as the starting point to create the gage specification. A variety of tolerance combinations (size and position) could be used in creating the gage. Only two alternatives are examined here. The gage tolerances are developed by first utilizing 10% of the part feature's positional tolerance to establish possible gage tolerances of size and position.

initial gage tolerance = part feature positional tolerance × 0.10

initial gage tolerance = 0.022 × 0.10 = 0.0022

A10.1.1   First Trial

The first trial will treat this value (0.0022) as the sum of the size and positional tolerances on the gage pin. Letting the first trial share the tolerance equally between size and position, the gage pin callout becomes

$$4X\ \varnothing.2500\ {}^{+.0011}_{-.0000}$$

$\oplus$	$\varnothing.0011$	A	B	C

At true position and using a $\varnothing.2500$ pin, the feature's virtual condition boundary is the same as the gage pin boundary; in theory, the $\varnothing.2500$ virtual condition hole will mate with a $\varnothing.2500$ gage pin.

However, if the LMC ($\varnothing.2500$) gage pin is $\varnothing.0011$ out of position, the following situation results (Figure A10-1):

Boundary of LMC pin ($\varnothing.2500$)	0.12500
Shift of pin center allowed by maximum positional error	-0.00055
Distance from true position to pin boundary	0.12445
Radius of LMC hole ($\varnothing.278$)	0.13900
Distance from true position to pin boundary	-0.12445
Difference between true position and hole center	0.01455

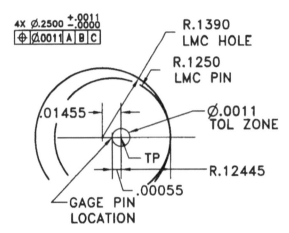

FIGURE A10-1   Functional gage with RFS positional tolerance, first trial.

The result, 0.01455, is the radial shift allowed to the center of an LMC hole. This shift is with respect to the true position location and can be converted into the diametral tolerance zone for comparison with the part specification. The resulting diametral zone is

$$\text{positional tolerance zone} = 0.01455 \times 2 = \varnothing.02910$$

When compared to the $\varnothing.028$ maximum positional tolerance the hole is allowed when it is fabricated at its LMC size, this preliminary gage design will allow nonconforming parts to be accepted. The gaging combination of an LMC pin at its maximum allowable positional error checking an LMC hole on the workpiece does not meet the absolute criterion.

A10.1.2   Second Trial

In an attempt to improve the design of the RFS gage and protect the specification at the extremes circumstances, the gage's inner boundary can be enlarged. Here 0.0011 has been added to the size specification of the gaging element. The new specification is

Checking the effects of this gage pin and feature combination, one can investigate

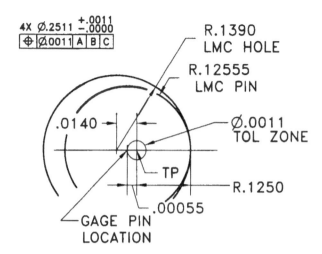

4X ⌀.2511 +.0011 −.0000

| ⊕ | ⌀.0011 | A | B | C |

FIGURE A10-2 Functional gage with RFS positional tolerance, second trial.

the results with the gage pin at LMC with maximum positional error shown in the diagram (Figure A10-2).

Boundary of LMC pin (⌀.2511)	0.12555
Shift of pin center allowed by maximum position error	−0.00055
Distance from true position to pin boundary	0.12500

Radius of LMC (⌀.278) hole	0.13900
Distance from true position to pin boundary	−0.12500
Difference between true position and hole center	0.01400

The result, 0.01400, is the radial shift allowed to the center of an LMC hole on the part being inspected. This shift is with respect to the true position location and can be converted into the diametral tolerance zone for comparison with the part specification. The resulting diametral zone is

positional tolerance zone = 0.01400 × 2 = ⌀.0280

When compared to the ⌀.028 maximum positional tolerance the part is allowed when it is fabricated at its LMC size, this second design will, in theory, accept parts within the design specification. However, it should be noted that this occurs only at the extreme inner boundary of the conformance region. Furthermore, this also requires that the gage be intentionally produced with a positional error that matches the maximum stated tolerance of position.

A different situation results if the gage is not produced with the intentional error mentioned, but is made with the LMC gage pin (⌀.2511) at true position.

A gage with an LMC gaging element at perfect location will encroach into the virtual condition boundary—now described with a diametral rather than radial value—of the workpiece (∅.2500) by ∅.0011. In addition, this encroachment could be as large as ∅.0022 if the pin is still at true position but fabricated at its MMC size. The result is that the RFS callout may reject more "good" parts than the "zero at LMC" gage design.

A10.2 MMC CALLOUT ON GAGE PIN

Another possible approach to the specification of gaging elements is to use traditional gagemaker's tolerances with MMC callouts. For this example, application of the 10% rule could lead to the following gage pin specification:

$$4X \; \varnothing.2500 \; {}^{+.0006}_{-.0000}$$

| ⊕ | ∅.0022Ⓜ | A | B | C |

A10.2.1 LMC Gage Pin at True Position

An LMC gage pin at true position is ∅.2500, the same as the virtual condition of the inspected feature. In theory, the LMC pin will match the virtual condition boundary of the workpiece and accept good parts while rejecting nonconforming parts.

A10.2.2 MMC Gage Pin at True Position

An MMC pin (∅.2506) at true position is ∅.0006 larger than the ∅.2500 virtual condition size of the mating hole on the workpiece. The result is that an MMC pin at true position could reject good parts between the workpiece virtual condition size of ∅.2500 and the MMC size of the pin, ∅.2506.

A10.2.3 Gage Pin at LMC and Maximum Positional Error

If the gage pin is fabricated at its LMC size and is out of position by the maximum positional tolerance allowed, the following situation (Figure A10-3) occurs:

Boundary of LMC pin (∅.2500)	0.12500
Shift of pin center allowed by maximum positional error	−0.00140
Distance from true position to pin boundary	0.12360
Radius of LMC (∅.278) hole on part	0.13900
Distance from true position to pin boundary	−0.12360
Difference between true position and hole center	0.01540

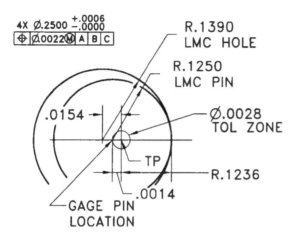

FIGURE A10-3   Functional gage with MMC positional tolerance.

The result, 0.01540, is the radial shift allowed to the center of an LMC hole on the part being inspected. This shift is with respect to the true position location and can be converted into the diametral tolerance zone for comparison with the part specification. In this situation, the resulting diametral zone is

positional tolerance zone $= 0.01540 \times 2 = \varnothing.0308$

When compared to the $\varnothing.0280$ maximum positional tolerance the LMC hole is allowed, this gage design would allow nonconforming parts to be accepted.

A10.3   ZERO POSITIONAL TOLERANCE AT MMC

The last approach to specifying gaging elements is to use traditional gagemaker's tolerances with MMC callouts but specify a zero positional tolerance. The difference between this approach and that of Sec. 10.6 is that manufacturing has the ability to partition the total feature tolerance of $\varnothing.0028$ in any manner—size or position. Application of the 10% rule leads to the following gage pin specification:

4X $\varnothing.2500$ $\begin{array}{c} +.0028 \\ -.0000 \end{array}$

| $\oplus$ | $\varnothing.0000$Ⓜ | A | B | C |

A10.3.1   LMC Gage Pin at True Position

The LMC size gage pin is $\varnothing.2500$, the same as the virtual condition of the inspected feature. If the LMC pin is located at true position, in theory it will match the virtual condition boundary of the workpiece and accept good parts while rejecting parts that are out of specification.

A10.3.2 MMC Gage Pin at True Position

An MMC pin (∅.2528) at true position is ∅.0028 larger than the ∅.2500 virtual condition size of the mating hole on the workpiece. The result is that an MMC pin at true position will reject good parts between the workpiece virtual condition size of ∅.2500 and the gage pin MMC size of ∅.2528.

A10.3.3 LMC Gage Pin and Maximum Positional Error

If the gage pin is fabricated at its LMC size and is out of position by the maximum positional tolerance allowed, the following situation (Figure A10-4) occurs:

Boundary of pin LMC (∅.2500)	0.12500
Shift of pin center allowed by maximum positional error	−0.00140
Distance from true position to pin boundary	0.12360
Radius of LMC hole	0.13900
Distance from true position to pin boundary	−0.12360
Difference between true position and hole center	0.01540

The result, 0.01540, is the radial shift the center of an LMC workpiece hole is allowed. This shift is with respect to the true position location and can be con-

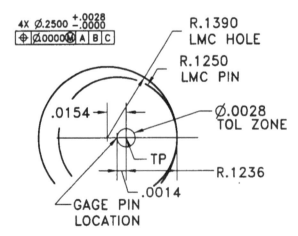

FIGURE **A10-4** Functional gage with zero MMC positional tolerance.

verted into the diametral tolerance zone for comparison with the part specification. In this situation, the resulting diametral zone is

positional tolerance zone = $0.01540 \times 2 = \varnothing.0308$

When compared to the $\varnothing.028$ maximum positional tolerance the LMC hole is allowed, this gage design would allow out-of-specification parts to be accepted. To reiterate, the difference between this approach and the MMC design of the preceding section is the increased ability to exchange size and positional tolerance to optimize the manufacturing process.

A10.4   DIAMETRAL ANALYSIS—ZERO AT MMC GAGE

Revisiting the example in Section A10.3, the following conventional analysis would result: The LMC hole remains the same.

The inner boundary of the gage pin becomes

gage pin's inner boundary = LMC − specified tolerance at MMC

−difference in size between MMC and LMC

gage pin's inner boundary = $\varnothing.2500 - \varnothing.0000 - \varnothing.0028 = \varnothing.2472$

maximum positional tolerance = LMC hole − gage pin's inner boundary

maximum positional tolerance = $\varnothing.2780 - \varnothing.2472 = \varnothing.0308$

This matches the values found in the earlier analysis. Figure A10-5 shows the effects of the MMC specification. Contrasting this with the LMC gage shows the additional shift allowed by the size specification of the gage pin. The result is that the acceptance region is enlarged when the fabricated pin departs from MMC size.

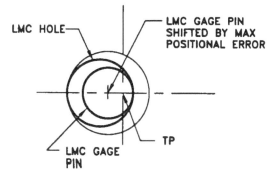

FIGURE **A10-5**   Functional gage analysis using diametral specifications.

11

Functional Inspection Techniques

11.1 INTRODUCTION

This chapter discusses inspection techniques that seek to duplicate, with varying levels of success, the information provided by hard gaging. These fall into three groups. First, surface-plate methods are discussed that may aid in developing the product definition if included as an element of the component design methodology. Additionally, machine-based techniques exist using CMMs and optical comparators—possibly involving surface-plate accessories and fixtures—to create the data set. Last, paper gaging, which continues to evolve into computer-based methods, can be used. The last technique (e.g., soft gaging), attempts to replicate hard gaging using data sets taken either by CMMs or by other inspection equipment.

Each of these tools has its own characteristic applications, and each has limitations that need to be kept in mind as the inspection process is designed. Arguably the most important concept used to understand the limitations of these techniques is measurement uncertainty. There is an unnerving tendency to attach low values of error (uncertainty) to any machine that digitally acquires data and manipulates it using a computer. Unfortunately, while such machines are of great benefit in the correct application, it is also easy to misapply or misuse them in ways that yield either unacceptable or unintended results.

11.2 FUNCTIONAL GAGING WITH SURFACE PLATES

Surface plates can serve as the base for specially constructed functional gages whose gaging elements are partially or completely simulated by standard surface-plate inspection accessories. These accessories have been referred to as "universal functional gages" by at least one source (Tandler, 2000). Such gages are carefully constructed setups of fixed geometry that resemble receiver gages and can be assembled with almost any degree of accuracy. They are clamped together with mechanical or magnetic force and can be calibrated in place when required.

As mentioned, the concurrent engineering team can use many of these techniques to simulate hard gaging. As such, they provide a reality check for the team even in the instance where a CMM will provide the production inspection capability.

Surface-plate gages can include wear allowances and can even be toleranced against the part if they are used as emergency measures until actual gages become available. Surface-plate gaging should be considered whenever any or all of the following conditions exist:

1. A number of parts, made from the same drawing, must be inspected.
2. The parts do not require variable inspection data; that is, simple Go or Not Go attributes are sufficient.
3. Schedules are tight and an in-process or acceptance gage is needed quickly.
4. Open-setup inspection of variables on the surface plate would be time-consuming because parts are difficult to set up or part surfaces are hard to contact reliably.
5. The parts do not lend themselves to paper gaging analysis.

With the advent of shop-floor CMMs and a significant reduction in their cost, functional gaging and functional inspection with surface plates may be less likely to be the preferred option. Yet they are still useful techniques in some applications. Manufacturing personnel who understand how these more traditional techniques may be applied to a given situation are more apt to apply coordinate metrology in ways that gather the required data for the decisions at hand. The following exposition serves to show how parts may be set up and data gathered, serving to expand understanding of the underlying processes and ensure that the correct information is acquired.

11.2.1 Gaging Positional Tolerances

As an example of gaging an MMC hole pattern, Figure 11-1 shows a part containing a pattern of clearance holes dimensioned from part datums A, B, and C.

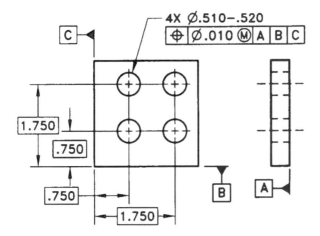

Figure 11-1 Drawing for a clearance hole pattern.

Each hole has a variable tolerance of position allowed by the MMC callout (0.010-in. tolerance for a ⌀.510 in., 0.017-in. tolerance for a ⌀.517 in., etc.).

The setup gage (Figure 11-2) is complex and requires careful planning. If the part is nonmagnetic, the gage can be held in place by magnetic force on a plate. The four pins are all ⌀.500 in. (0.510-in. hole at MMC minus the 0.010-

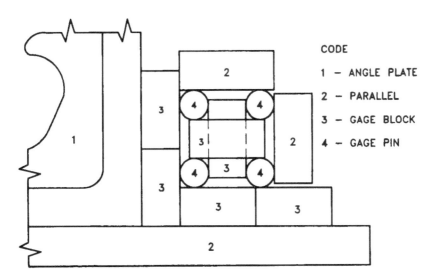

Figure 11-2 Surface-plate hole pattern gage (plan view).

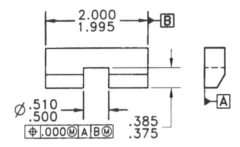

FIGURE **11-3**   MMC specification for symmetry.

in. tolerance allowed at MMC), and their basic center locations are 0.750 in. and 1.750 in. from the angle plate and the parallels. The part is acceptable if the holes fit over the gage pins and surfaces B and C contact the angle plate and parallels. It should be possible to position the primary datum feature (A) of the part parallel to the surface plate, which supports the surface-plate gage items in the plan view. The ∅.510–.520-in. hole size limits must be checked separately to complete the inspection.

Figure 11-3 shows a part with a symmetry requirement (specified by a positional tolerance) that is difficult to inspect by conventional means. The gage stack shown in Figure 11-4 is a true Go functional gage. The gage setup has a 2.000-in. gap between the outer gage blocks to accommodate part datum feature B at MMC (2.000 in. on the finished part). It also has a 0.500-in. slot gage— the slot is allowed no symmetry tolerance when it is 0.500 in., and up to 0.010-

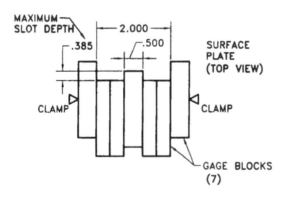

NOTE: ALL SEVEN BLOCKS MUST BE
CLAMPED TO MAINTAIN GAGE INTEGRITY.

FIGURE **11-4**   Surface-plate setup for gaging symmetry.

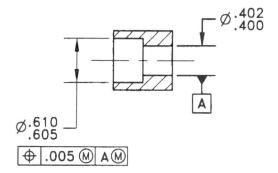

FIGURE 11-5   Specification for coaxiality (MMC datum).

in. symmetry tolerance when it is 0.510 in.). The entire gage stack is placed on a surface plate, which must contact part datum feature A during gaging. If the part enters the gage, the inspector need only inspect the 0.510-in. maximum slot width and the 1.995-in. minimum datum feature B width to complete the inspection.

Figure 11-5 shows another part that is difficult to inspect. The single functional gage setup shown in Figure 11-6 will relieve the inspector of several calculations and individual setups involving the actual size of datum feature A. The ∅.600-in. gage pin (0.605-in. MMC hole minus the 0.005-in. positional tolerance allowed at MMC) and the ∅.400-in. pin (the MMC size of the datum hole) should be coaxial. If both pins enter and touch bottom in the holes, all the inspector needs to do to complete the inspection is to check the 0.605-, 0.610-, and 0.402-in. limits. The part should not be forced against either V block while being inspected.

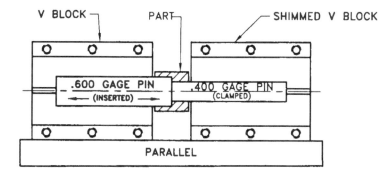

FIGURE 11-6   Surface-plate setup for gaging coaxiality (plan view).

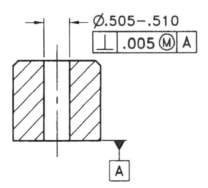

FIGURE 11-7   MMC specification for perpendicularity.

11.2.2   Gaging Form and Orientation Tolerances

Perpendicularity.   Figure 11-7 illustrates a part to be measured against a variable perpendicularity tolerance. When the hole is finished to ∅.505 in., the perpendicularity tolerance is 0.005 in; when the hole is finished to a ∅.510 in., the perpendicularity tolerance increases to 0.010 in. All finished hole sizes between ∅.505–.510 have corresponding perpendicularity tolerances.

Figure 11-8 shows the functional gage setup for measuring variable perpendicularity, consisting of a ∅.500-in. gage pin (∅.505-in. MMC hole minus the

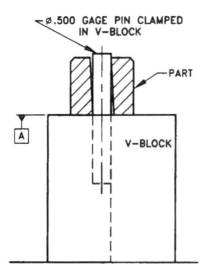

FIGURE 11-8   Setup for gaging perpendicularity.

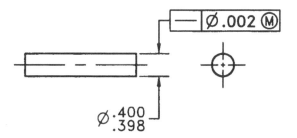

FIGURE 11-9 MMC specification of straightness.

Ø.005-in. perpendicularity tolerance allowed at MMC) clamped in a V block. If the part fits over the pin and datum feature A makes flush contact with the V block, the part is acceptable, providing the hole diameter is within the 0.505- to 0.510-in. limits.

Straightness. Figure 11-9 shows a pin whose axis must be straight within a Ø.002-in. zone when it is at MMC (Ø.400 in.) and can deviate within a straightness tolerance of Ø.004 in. when it is at Ø.398 in. Instead of measuring each pin to its exact diameter and then computing the resulting straightness tolerance (that is, a Ø.3992-in. pin is allowed a 0.0028-in. tolerance), the inspector can inspect a series of these parts more rapidly with the setup shown in Figure 11-10.

The surface-plate gage in Figure 11-10 has a 0.402-in. opening (0.400-in. plus 0.002-in. straightness tolerance at MMC). The pin need only roll under the bridge (making a complete revolution) to be acceptable, provided other measurements have shown that it is no larger at any cross-section than Ø.400 in. and no smaller than Ø.398 in.

Profile. Figure 11-11 shows a profile tolerance of 0.010 in. normal to the basic contour. (The series of coordinate profile-locating dimensions have been omitted for clarity.) The inspection of such a requirement is time-consuming with normal setup procedures. Figure 11-12 shows a Go gage setup for inspecting contour tolerances consisting of standard surface-plate equipment. The pin sizes

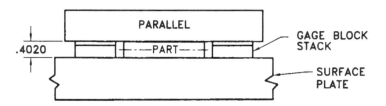

FIGURE 11-10 Surface-plate setup for gaging straightness.

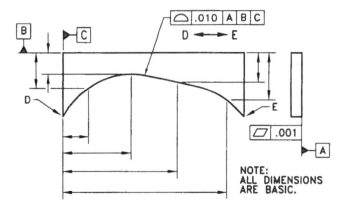

FIGURE 11-11   Contour tolerance specification.

are determined by drawing a layout of the contour and fitting a series of circles (representing gage pins) to the contour so that the circles contact the maximum (Go) profile. This type of gage can rapidly monitor (as contrasted with verifying conformance) production of parts too large for optical projectors.

11.3 FUNCTIONAL GAGING WITH COORDINATE MEASURING MACHINES

Coordinate measuring machines can be considered to perform three possible tasks: reporting actual values; manufacturing process feedback; and compliance

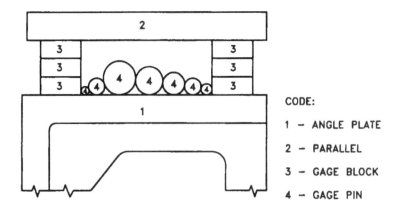

CODE:

1 — ANGLE PLATE

2 — PARALLEL

3 — GAGE BLOCK

4 — GAGE PIN

FIGURE 11-12   Setup for gaging contour (plan view).

testing. The first two uses are arguably the machines' strong suit. Compliance testing requires more care and ingenuity on the parts of the planner and the programmer to ensure the results are generated at an acceptable level of uncertainty.

Coordinate measuring machines can emulate functional gages through their software. Hypothetically, a more extreme view might see them converted to universal functional gaging devices by adding a precision chuck to permit the use of standard gage pins and bushings instead of probes. While this emphatically would not be done, theoretical consideration of this technique serves to illustrate what the CMM is attempting to simulate through software. Ultimately, the mental exercise allows the inspection planner to understand both the advantages and the disadvantages of applying the CMM in functional gaging situations. It also helps guide programming decisions when constructing the virtual geometry of the inspection model.

11.3.1   Functional CMM Programming

When emulating functional gaging using a CMM, a series of steps (after Tandler, 2000) must be explicitly incorporated in the inspection process. Assuming a measurement process plan exists, the first step involves acquiring a raw data set. In the case of CMMs, this may involve the use of a probe with a finite tip radius. The raw data is collected as the location of probe center points that require correction for the probe's finite radius. The tip radius will introduce mechanical filtering effects along with contributions to measurement uncertainty. Additionally, such a method raises questions relating to which surface points the probe actually contacted on the workpiece. Uncertainty arises from the inability to calculate the specific correction factors without full knowledge of the geometry of the workpiece. This does not substantially differ from other measurement processes involving the use of a finite radius on a probe.

Initial data collected is used to establish an alignment between the machine and the workpiece. The alignment locates the part with respect to the machine's coordinate system. This accomplishes with software what would be a time-consuming process if the part needed to be physically aligned within the machine. In its simplest form, the probed points establish the primary, secondary, and tertiary planes and create the alignment.

To create the base alignment, subsets of the corrected data are used to construct substitute geometry. The construction methods involve a series of decisions by the programmer that deal with geometry in a virtual, three-dimensional space where different geometry construction techniques—including number of points and local point densities—might be used and many candidate projection planes exist. Creating the desired geometry requires selecting the appropriate planes upon which to project geometry and choosing the appropriate commands from the software command set to create the geometric entity.

As pointed out by Tandler, in CMM programming, "The more imperfect the geometry, the more difficult its assessment." Sheetmetal components are one excellent example of this type of geometry. Particularly, with highly flexible sheetmetal parts, determination of datum feature location and controlled features can be challenging undertakings for the metrologist. Pitfalls that await the programmer might include the construction of geometric entities such as circles and cylinders. In particular, as form errors become significant contributors to the actual geometry, assessing the characteristics of the cylinder (e.g., the diameter and axis of a cylinder) begins to test the ingenuity of the programmer. This may involve a multitude of decisions concerning the geometry construction tools to be used and the coordinate systems within which these tools will be applied. Projection of the probed points onto an incorrect projection plane (i.e., into an incorrect DRF) can significantly alter geometry and yield invalid reported values. A specific example would be fitting a circle to series of points taken from a "perfect" cylinder. Projected onto an incorrect plane, the perfect geometry falsely reports an out-of-roundness condition.

Following the measurement plan, the substitute features that have been created are now used to construct DRFs. At this point it is assumed that the product definition has been sufficiently analyzed so that the programmer can arrive at valid interpretations of the necessary coordinate systems (i.e., DRFs) to support inspection. The DRF may be equivalent to the single functional frame touted throughout this book or may include multiple DRFs that can occur in extremely complex product definitions. The results of the analysis provide the information—the DRFs—needed to construct the virtual equivalent of datum feature simulators and incorporate these into what can be described as a virtual functional fixture. This software-based fixture provides the coordinate system necessary to inspect the controlled features.

In essence, the programming process for the CMM requires that the measurement process planner or the CMM programmer design a software equivalent of the functional gage. When the programmer creates the DRFs within the software, he or she mimics the design and manufacture of the fixture elements of a functional gage. The result is the functional coordinate system required to assess conformance of the controlled features.

The remainder of the process involves completing the inspection of the controlled features. This requires probing additional points to create substitute features as proxies for the actual controlled features. It may also include the construction of additional coordinate systems when multiple DRFs are found in the product definition or prove necessary for inspection purposes.

11.3.2   Hypothetical Conversion

Figure 11-13 shows how this imaginary exercise of converting a CMM into a functional gage might be put into practice. The calibrated precision chuck forms

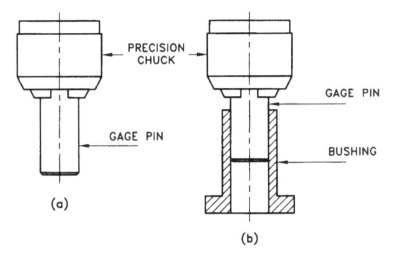

FIGURE 11-13 Hypothetical conversion of coordinate measuring machine to functional gage.

the basis for this conversion. Once the precision chuck is in place, standard gage pins available in 0.0001-in. increments could be used [Figure 11-13(a)]. The pins could be slightly oversized to reflect wear allowances and gage tolerances. Alternatively, bushings might be used, still based on the installation of a precision chuck and gage pins to hold the bushings [see Figure 11-13(b)].

11.3.3 Examples

This imagined conversion of the CMM, modified for functional gaging, can be used to check size, perpendicularity, and location of holes, pins, and tapped holes. In the following discussions, there should be consideration of measurement uncertainty as it relates to simultaneous versus separate requirements. The software-based functional gaging capability does not completely simulate the hard gage; in particular, the effects of the physical assembly process where a number of part features are assembled simultaneously should be considered. Obviously, the hypothetical examples assume manual control of the CMM.

11.3.3.1 Holes

Perpendicularity. Figure 11-14 shows the method of checking hole perpendicularity. Using the actual mating size of the hole—which includes a specification of a basic orientation for this application—to determine pin size, the gaging element mimics the similar element that would be found in the functional gage and used to check the hole for compliance. If gage pins are interchanged until the largest pin is selected that just fits in the bore, one is said to be inspecting

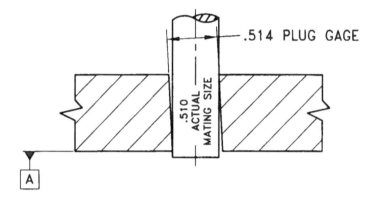

.510 ACTUAL MATING SIZE

.514 PLUG GAGE

A

FIGURE 11-14 Proposed check of actual mating size at basic orientation.

the orientation-constrained actual mating size of the feature. In this example, the hole accepts a Ø.514-in. plug gage (regardless of orientation), which means that it has at least a Ø.514 in.

However, if the hole is not perpendicular to primary datum surface A, it will not accept a Ø.514-in. mating part feature. (This can be readily determined by inserting the Ø.514-in. plug gage in the precision chuck mounted on the machine spindle and attempting to push the gage through the hole.) Note in Figure 11-14 that the largest perpendicular pin that will pass through the hole has a Ø.510 in. This is the virtual size of the hole and means that the hole axis is out of perpendicularity to the primary datum surface by 0.004 in. (0.514 − 0.510 in.).

Location. Figure 11-15 shows a straight pin being used to check the location of a positionally toleranced hole. Suppose the callout for the hole is Ø.510–.530, with a Ø.010-in. positional tolerance at MMC. The size of the straight pin to be used for gaging location is determined by subtracting the 0.010-in. MMC positional tolerance from the MMC diameter of the hole, which is 0.510 in. Thus, Ø.510 − Ø.010 = Ø.500-in. pin.

The spindle is positioned at the basic hole location by the machine (datum feature A should be perpendicular to the machine spindle), and the pin is passed through the hole. This procedure is repeated at each hole location. Separate gaging operations to determine hole size with Ø.510-in. Go and Ø.530-in. Not Go gages complete the inspection operation.

If the callout for the hole were Ø.500–.530, with zero positional tolerance at MMC, the same Ø.500-in. pin (Ø.500 − Ø.000 in.) would be used to gage the hole location. However, the zero MMC tolerancing specification would eliminate the need for a separate Go gage (Ø.510 in.) to gage minimum hole size, because the Go gage is incorporated in the Ø.500-in. straight pin. Thus, location

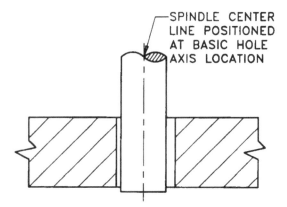

FIGURE **11-15** Gaging a hole location.

and minimum size are gaged at the same time. A separate Ø.530-in. Not Go gage is still required to gage maximum size.

11.3.3.2 Pins

Figure 11-16 shows a bushing used to check the location and size of a dowel pin. Suppose the callout for the dowel pin is Ø.4996–.5000, with a Ø.010-in. positional tolerance at MMC. The inside diameter of the bushing to be used for gaging is determined by adding the 0.010-in. MMC positional tolerance to the MMC diameter of the hole (0.500 in.). The result is Ø.500 + Ø.010 = Ø.510-in. bushing internal diameter.

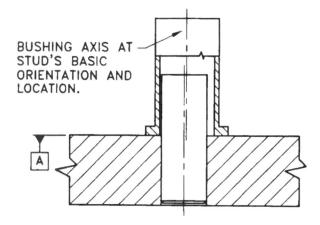

FIGURE **11-16** Gaging a stud for location.

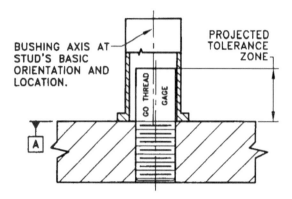

BUSHING AXIS AT
STUD'S BASIC
ORIENTATION AND
LOCATION.

PROJECTED
TOLERANCE
ZONE

GO THREAD GAGE

A

FIGURE 11-17   Gaging a tapped hole for location.

The spindle is positioned at the basic pin location (datum feature A should be perpendicular to the machine spindle) and the bushing is lowered over the pin. The bushing should touch the top surface of the part. The operation is repeated at each pin location.

11.3.3.3   Tapped Holes

Figure 11-17 illustrates a bushing used to check the location and size of a tapped hole. Suppose the callout for the tapped hole is ¹/₂-13 UNC-2B, with a positional tolerance of ∅.010 in. at MMC and a projected tolerance zone of 0.750 in., which is the maximum thickness of the mating part. The inside diameter of the bushing is determined by adding the ∅.010-in. positional tolerance to the MMC diameter of the bolt shank that will fit the tapped hole. Thus, ∅.500 + ∅.010 = ∅.510-in. bushing ID.

A ∅.500-in. Go thread gage with a ∅.500-in. shank is inserted into the tapped hole. The shank should project 0.750 in. above part surface A. Datum surface A should, of course, be perpendicular to the machine spindle. The spindle is positioned at the basic tapped hole location and the bushing lowered over the thread gage shank. The operation is repeated at each tapped hole location.

For low-volume work, the Go thread gage can be moved to each tapped hole and inserted prior to the inspection. Parts should not be repositioned during the above gaging operations, as this would destroy pattern integrity.

11.4   FUNCTIONAL GAGING WITH OPTICAL COMPARATORS

Optical comparators can simulate functional gages by using specially constructed chart gages, which are two-dimensional simulations of three-dimensional receiver gages resembling the most critical mating part. Thus, optical chart gages can be used to directly gage positional tolerances modified with MMC specifications—and profile tolerances that define size limits. This occurs because any fixed set of circles on a chart is representative of the fixed gage pins on a receiver gage and will automatically allow tolerances to vary as hole sizes vary.

11.4.1   Applications

Small, relatively thin parts are usually quite suitable for optical measurement because they may be magnified numerous times. Most optical projectors have available magnifications of 10, 20, 50, or 100, with some machines incorporating zoom capability. This phantom-gaging technique is particularly applicable to parts entirely defined with profile tolerance zones because these boundaries describe the exact optical projector chart gage outlines.

The following description illustrates the use of this technique:

1. Draw the part to scale, using nominal (basic) dimensions.
2. Superimpose the tolerance zones on the basic profile using either unilateral or bilateral zones as required. These zones define the limits of size and the location of the part features. See Figure 11-18(a).
3. The lines representing the tolerance zones [the phantom-gage outlines in Figure 11-18(a)] are the chart gage that will be placed on the viewing screen of an optical projector. The final part shadow, when magnified to the same scale as the chart gage, must lie within the chart gage tolerance zones. Figure 11-18(b) illustrates the functional gage design demanded by the part optically gaged by the phantom outlines.

The minimum distance between chart gage lines should be great enough for the inspector to conveniently resolve and never be closer than 0.020 in. Thus, a 0.002-in. tolerance zone width must be magnified at least 10 times. This rule is a convenient guide for determining a minimum chart gage scale.

Datum targets can be conveniently and directly represented on a chart gage, using small circles for dowel locators, dashed lines for parallel bars, and so on. The primary datum feature is aligned with appropriate staging fixtures so that it is perpendicular to the collimated light beam.

11.4.2   Profile Tolerancing

As mentioned, when profile tolerancing is used, the tolerance zones are the chart gage lines, and a pictorial tolerancing concept is used to describe part form and size limits. Figure 11-19 shows how both profile and positional tolerancing could be employed to define such a part. This concept is discussed in some detail in the section on phantom-gage dimensioning in Chapter 6.

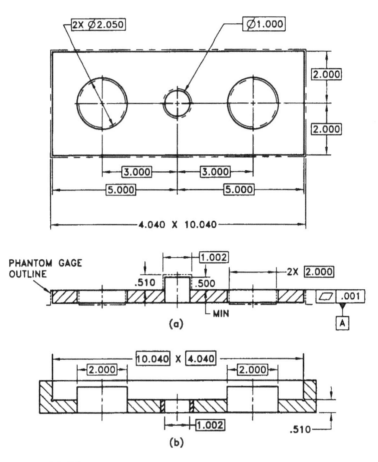

FIGURE 11-18 Part entirely defined with phantom-line tolerance zones and functional gage.

The primary datum is feature A, identified by the chamfered edge, and is staged so that it is perpendicular to the collimated light beam. The secondary datum feature is the outside contour. This entire contour of the part is then controlled using a profile tolerance related to the primary datum. The two Ø.105–.110 holes at MMC are then controlled to DRF consisting of A and the profile. The part cannot be rotated or translated for fit if finished to the MMC size of the datum feature, because it cannot be moved when its shadow is compared to the MMC contour on the chart gage [see Figure 11-20(a)]. If the B datum feature is finished smaller than its MMC size, the part can be adjusted to allow the two circles representing gage pins to move within the part holes.

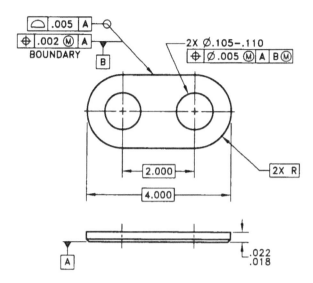

FIGURE 11-19 Profile and MMC positional tolerance.

The two circles are each Ø.100 in., multiplied by the chart gage magnification factor. The Ø.100 in. is calculated as follows:

Ø.105, diameter at MMC

− Ø.005, positional tolerance at MMC

Ø.100, chart gage diameters before scale factor

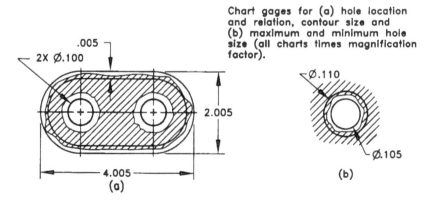

FIGURE 11-20 Chart gages for profile tolerancing with MMC positional tolerancing.

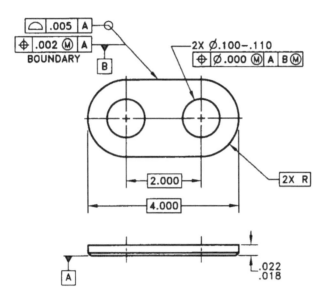

FIGURE 11-21   Profile and zero MMC positional tolerancing.

Figure 11-20(a) shows that the chart gage simultaneously determines if the exterior profile complies with the basic profile contained in the product definition. This would be required for control of the profile to datum A.

Figure 11-20(b) shows the separate chart gage, which is required to check hole size. This chart gage is comparable to separate Go and Not Go plug gages and is used to check each hole size in the pattern independently.

Figure 11-21 shows the part of Figure 11-19 redefined with zero MMC positional tolerancing, which eliminates the separate Go check. The chart gage for Figure 11-21 is contained in Figure 11-22. Note that the separate inside line of Figure 11-20(b) has been eliminated in Figure 11-22(b); the Go configuration is included in the Ø.100-in. circles of Figure 11-22(a).

Profile tolerancing lends itself to inspecting stamped parts, particularly those produced by a single operation on automatic equipment. Part stamping dies can be designed so that, as they wear, the changes in size of the parts produced will stay within the tolerance zones during the life of the dies.

11.5   PAPER LAYOUT GAGING

Paper layout gaging is a direct and inexpensive technique for making an immediate functional check of inspection results, permitting the adjustments possible with functional three-dimensional receiver gaging. It provides a method of de-

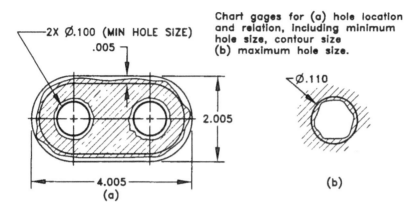

Chart gages for (a) hole location and relation, including minimum hole size, contour size (b) maximum hole size.

FIGURE 11-22   Chart gages for profile tolerancing with zero MMC positional tolerancing.

termining if a part can be reworked and, if so, the most economical rework required. It is also a useful way to evaluate tooling, indicating the adjustments required to produce an acceptable product. Moreover, paper layout gages neither wear out nor require storage space, as do receiver gages.

Conventional dimensioning and tolerancing techniques usually employ tighter tolerances than necessary to make sure that manufacturing stays within required limits. Gaging the first tool-made sample part in a production run with a paper gage can immediately tell a manufacturer how well the tooling will meet the design specifications during actual production. This might encourage the manufacturer to relax his or her tolerance and size limits or establish new tooling "nominals," thus increasing potential acceptance rates.

11.5.1   Application

The first step in applying the "paper gaging" techniques is to decide when to use it. An examination of the inspection report will yield certain essential information about design specifications and inspection procedures, indicating whether or not paper gaging is necessary or feasible.

A part with coordinate datum dimensions, all originating from a single specified DRF, does not require paper gaging, regardless of the tolerancing method, because the part features are fixed in relation to the datum reference frame (see Figure 11-23). Conformance can be checked mathematically by using the inspection report rather than resorting to a graphical technique that adds additional and unnecessary steps to the inspection process.

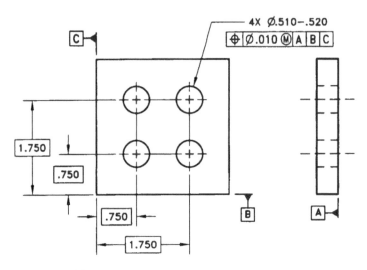

FIGURE 11-23   Parts that should not be paper-gaged.

Some parts can be paper-gaged, but the procedure is more trouble than it is worth. It is better and easier to analyze the inspection results directly. A prime example of this is inspecting a part with coaxial requirements.

Figure 11-24 shows a drawing of a part that must be coaxial about a datum axis. Part coaxiality, if specified as a positional tolerance, can be partially checked by rotating the part about its datum axis and measuring surface runout with an indicator. This may be a feasible inspection method in the situation where the

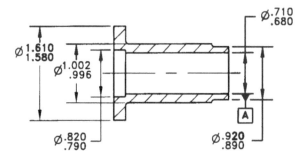

FIGURE 11-24   Circular runout to control coaxiality.

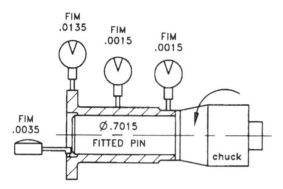

FIGURE 11-25   Inspection setup for coaxiality example in Figure 11-24.

process adequately controls form such that the remaining form errors do not influence measurement uncertainty.

There are numerous ways to determine the axis of a datum feature. In this functional case, a gage pin, ∅.7015 in., just fits datum feature A. The setup, using this pin inserted into a precision chuck, is shown in Figure 11-25.

Note that the FIM (full indicator movement) checks in Table 11-1 can be deceiving, because they include elements of runout, out-of-roundness, measurement axis error, and so on. A detailed inspection of form should be made along with all FIM measurements if eccentricity is to be segregated.

11.5.2 Parts That Can Be Paper-Gaged

Parts with independent patterns of axial features (holes or pins, for example) related only to a primary datum, which would ordinarily be gaged with a feature relation receiver gage, can be effectively paper-gaged. These situations arise when patterns of features are used as datums or for the pattern control that occurs with composite positional callouts.

TABLE 11-1   Inspection Results

Measured feature diameters (in.)		Tolerance (in.)		Runout tolerances (in.)	FIM
0.792	+	0.0015	+	0.002 (0.792 − 0.790)	0.0033
1.598	+	0.0015	+	0.012 (1.610 = 1.598)	0.0133
1.002	+	0.0015	+	0.000 (1.002 = MMC)	0.0013
0.902	+	0.0015	+	0.000 (0.920 = MMC)	0.0013

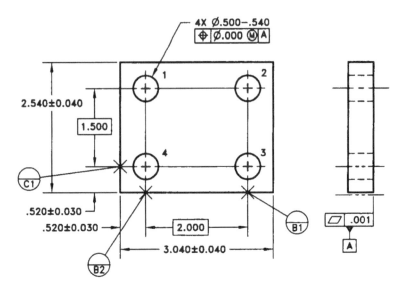

FIGURE 11-26 Independent hole pattern dimensioning.

Dimensioning to allow hole pattern independence on a part is shown in Figure 11-26. This type of part creates problems when the gage designer attempts to create a design. It would not be unusual to see that hole 4 is erroneously considered to be a datum. Many gage designers might therefore gage this hole with a tapered pin, which unduly restricts the relationship of hole 4 to the other holes in the pattern.

11.5.3 Paper Gaging Procedure

Separate layouts are made for a pattern of part features: one of measured axis locations, and the other of positional tolerance zones. The tolerance-zone layout is superimposed over the measured axis layout and rotated and translated to determine if any single orientation allows all plotted axis points to fall within their respective tolerance zones. The procedure is analogous to drawing an extremely large part exactly as described on the inspection report and comparing it to another equally large drawing of the gage to see if the part fits. Obviously, this particular procedure is impractical because the scale factor required to make the tolerances visible would create drawings to fill a room.

The part shown in Figure 11-26, which will be used to illustrate the procedure, can be inspected using two outside peripheral surfaces or any two holes as secondary and tertiary datum features. In any case, datum surface A is used as the primary datum. The results of one inspection procedure are shown in Table

8-3 (reproduced here as Table 11-2), based on Figure 11-26. Two outside surfaces were used as secondary and tertiary datums, and the part was set up as shown in Figure 11-27. The letter "A" modifying a measurement in Table 11-2 merely indicates that this was the coordinate location of a hole center point nearest datum surface A.

Scaling Tolerances. Tolerances are generally cited with an order of magnitude in thousandths of an inch. For this reason, tolerance zones and measured axis variations must be scaled up so that they can be seen and accurately plotted and evaluated.

Scaling Dimensions. If dimensions are scaled up by the same factor as tolerances, the layout will be too large to handle. The same scale factor is therefore not used for dimensions unless the actual paper gaging operation indicates a marginal part.

Format. Experience dictates that the layout format should resemble the geometry of the part. Prepared plots of coaxial circles about a common axis are used, but this technique can lead to confusion if not carefully plotted and evaluated.

Material. Plastic materials such as Mylar can be used in place of paper for layouts. Mylar is somewhat more stable than paper and is reusable, since pencil lines on it can be easily erased. Several Mylar sheets and a grid can be used over and over again, with copies of each gage made as a permanent record.

11.5.4   Inspection Results Layout

The measured locations of points at both ends of each feature axis are plotted in relation to the x and y-axes established by the datums specified on the inspection report. The two perpendicularity points plotted for each hole axis can be joined by a line to indicate that they are points for the axis of one hole. This procedure creates the effect of a three-dimensional gage.

Grid Plot. Each square in a preprinted metric grid is assigned a value, perhaps 0.001 in. Because each square is 1 mm or 0.039 in. on a side, this would give a scale factor of 39 to 1. A piece of transparent sketch paper is laid over the grid, fixed so that it will not shift, and the basic locations of each axis plotted with a compass point for accuracy. As an aid to quick identification of part geometry, the center lines may be joined in a rectangular grid pattern.

The difference between the basic axis location and each measured axis point location is translated into metric grid squares and counted off in the x and y directions from each basic axis location plotted on the layout.

TABLE 11-2 Sample Inspection Report

Inspection Report

Figure: Datum. Setup: Surface A and two additional sides. Datum targets used are marked on part.

Item no.	Specified dimension, tolerance, etc.				Actual dimension as checked					
	x	y	Tolerance diameter at MMC	Size	x	y	Tolerance (actual)	Tolerance diameter allowed by MMC	Size	Results
1	0.520	2.020 (0.520 + 1.500)	0.000	0.500 / 0.540	0.538 0.540(A)	2.033 2.030(A)	See paper gage[1]	0.030	0.530	OK
2	2.520 (0.520 + 2.000)	2.020 (0.520 + 1.500)	0.000	0.500 / 0.540	2.548 2.545(A)	2.000 2.000(A)	See paper gage	0.035	0.535	OK
3	2.520 (0.520 + 2.000)	2.020 (0.520 + 1.500)	0.000	0.500 / 0.540	2.522 2.522(A)	0.497 0.497(A)	See paper gage	0.020	0.520	OK
4	0.520	0.520	0.000	0.500 / 0.540	0.505 0.510(A)	0.525 0.525(A)	See paper gage	0.025	0.525	OK

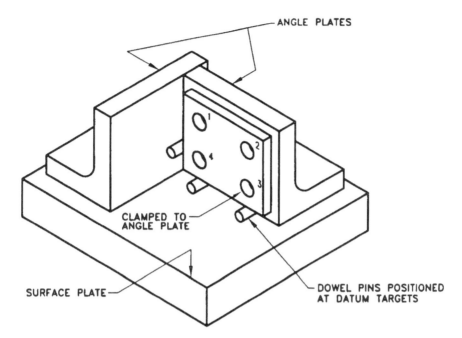

ANGLE PLATES

○¹
○²
○⁴
○³

CLAMPED TO
ANGLE PLATE

SURFACE PLATE

DOWEL PINS POSITIONED
AT DATUM TARGETS

FIGURE **11-27**   Setup to inspect part in Figure 11-26.

Figure 11-28 shows the inspection plot for Figure 11-26 and Table 11-2.

Drafted Plot.   The basic angle axis locations of a bolt circle or similar simple circular pattern can be laid out with drafting instruments or plotted with a CAD system. The basic locations of complicated or rectangular patterns can be handled in the same manner.

The measured axis locations are readily plotted from each basic location. An appropriate scale factor for plotting the measurements is selected, perhaps for each 0.001 in. of factual variation.

11.5.5   Tolerance Layout

It is important that the basic axis locations of the inspection results layout be plotted identically on the tolerance layout, which is the second layout required in paper gaging. This can be done by redrafting, or laying the tolerance layout sheet (or CAD layer) over the inspection results layout and transferring the basic locations.

The positional tolerance-zone diameters taken from the inspection report are scaled up by the same factor selected for plotting the measured axis locations and then drawn in.

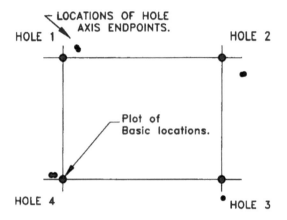

FIGURE 11-28 Plot of hole center deviations from inspection report.

A tolerance layout for Figure 11-26 and Table 11-2 is shown in Figure 11-29. The positional tolerance zones are plotted directly from the "Tolerance diameter allowed by MMC" column on the inspection report.

11.5.6 Combining Layouts

Because the sizes of the tolerance zones vary, the results of paper gaging will be entirely nullified if the tolerance zones are not placed over their corresponding

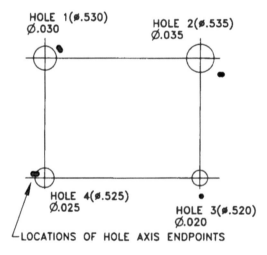

FIGURE 11-29 Plot of positional tolerance zones.

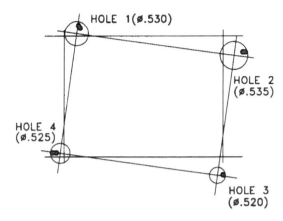

FIGURE 11-30 Completed paper gage.

axis locations. It is thus advisable to mark or number each hole location on both layouts of a complex pattern, particularly if the pattern is symmetrical.

The completed tolerance layout is placed over the inspection results layout and slightly shifted and/or rotated if necessary to determine if all sets of measured axis location points fall simultaneously within their respective tolerance zones. This graphically demonstrates whether or not the part being gaged will assemble with its mating component.

Figure 11-30 shows the completed paper gage for the part of Figure 11-26 and Table 11-2. The tolerance layout has been rotated and shifted, much the same as a receiver gage or mating part would be adjusted for assembly. All four sets of axis points (top and bottom readings) fall within their respective tolerance zones, demonstrating that the part is acceptable.

Measurements taken from any setup used for the part in Figure 11-26 will result in plots of hole centers and tolerance zones that have the same relationship to each other. The tolerance-zone plot (Figure 11-29) is uniformly the same, and the finished hole sizes and their perpendicularity remain constant for the same part. Thus, identical results are obtained with paper gaging.

11.5.7 Allowance Factors

The overall accuracy and validity of paper gages are affected by such factors as the accuracy of the layouts themselves, the accuracy of inspection measurements, and the completeness of inspection results. Also, allowances must sometimes be made for gage tolerances and gage wear if paper gaging is to be used in conjunction with functional receiver gages. Reducing the diameters of the tolerance zones on the paper gage layout by a fixed percentage can serve as an overall safety factor to compensate for inaccuracies and gage allowances.

Layout Accuracy. The material on which the layouts are made introduces error equivalent to its coefficient of expansion and contraction. If the same material is used for both inspection and tolerance layouts, this type of error cancels out. The grid used for a plot can introduce error if it is not a perfect grid (a printed or photocopied grid can be assumed to be somewhat imperfect). Using the same grid in an identical manner for making both layouts cancels out this error. There is likely to be at least a 0.010-in. error in positioning lines, points, and diameters. This error is directly minimized by the scale factor selected; thus,

$$\frac{\text{positioning error}}{\text{scale factor}} = \text{actual error}$$

In making the layouts, a 0.010-in. width of pencil line, with a 100 to 1 scale factor, can cause a 0.0001-in. error (0.010/100), 0.00005 in. on each side of the line. This error can be minimized by working to one side of the line.

Inspection Measurements. Obviously, a certain amount of uncertainty is inherent in inspection measurements. However, since open-setup inspection is used to calibrate tooling, fixtures, and gages, it can be assumed that carefully made measurements will not produce an error factor greater than 5% in the data on an inspection report.

Incomplete Inspection Results. Sometimes an inspection report is incomplete in that it does not contain information about setup or hole perpendicularity, yet a paper gage must be made from it. Therefore, some reduction in the size of the tolerance zones can be made to compensate for the uncertainty of incomplete inspection results. This reduction may be as much as 25% and not be unreasonable. Such a reduction is useful when the inspection report contains an uncertainty factor, as previously discussed, since the reduction can be directly applied to modify the tolerance-zone layout.

Gage Allowances. Reducing the size of each tolerance zone can also serve to include allowances for gage tolerances and wear. This is important if the paper gage is intended to accept or reject to the same degree as a receiver gage. Such would be the case when paper gages are used for in-process checks or at the beginning of a tool run when receiver gages are not yet available but will be used.

No allowance should be made on the paper gage for gage tolerance or wear if no receiver gage will be used for the part(s) being inspected. This would only lead to rejection of otherwise acceptable parts. Also, no reduction of tolerance-zone size is necessary to accommodate virtual size; the plotting of both ends of the axis to establish perpendicularity automatically takes care of this, because both points must fall within the tolerance zone. This also gives the paper gage a three-dimensional effect.

Gage Policies. The question of allowing for gage tolerances and wear brings up the problem of differing gaging policies. Currently, no standard presents an unequivocal interpretation of Go and Not Go gage tolerances, and there is no commonly accepted standard percentage of wear to be allowed before a gage is taken out of service to be reconditioned. It is extremely important for users and suppliers (or design, manufacturing, and inspection departments) to arrive at a common understanding on these matters before production starts.

11.5.8 Analyzing Results

In analyzing a completed paper gage, the true power of this technique becomes clear. Not only does it indicate functional acceptance, but it also can be used as the basis for determining the feasibility of reworking a part, the nature of the rework, and what tooling changes might be required to bring parts into acceptance. A series of paper gages made during a production run can be used to monitor the rate of tool wear and to predict accurately when tooling should be replaced or reworked.

Rework Determination. Paper gage analysis, including scaling plotted axis locations in relation to the perimeter of their respective tolerance zones, can be of great value in determining rework.

Suppose the hole centers lie outside their respective tolerance zones; if one or more tolerance zones can be increased in diameter (which is accomplished by reaming out the corresponding holes to a larger diameter), the part can be accepted. Measurement of the paper gage relationships between axis locations and tolerance-zone perimeters can be used to determine the amount of rework and when and where it is required.

Whoever makes the paper gage can indicate on the inspection report the rework required to bring the part into acceptance and indicate a provisional "OK" in the results column for the particular feature or features to be reworked.

Ordinarily the inspector does not judge acceptance of a part; he or she merely records inspection measurements on the report, and the designer determines acceptance. However, if the inspector makes the paper gage, he or she can indicate the required rework on the report. Of course, it is necessary that the part be reinspected and perhaps paper-gaged again after rework to make sure it is then acceptable.

Tooling Check. Succeeding parts made on the same tooling will look much the same as the first part, with only minor variations. Thus, necessary changes in tooling can be determined on the basis of one paper gaging operation, and perhaps additional perfunctory checks made during production. Paper gages can also be made on a regular basis to conveniently monitor tool wear so that the frequency and occasion for change and rework of tooling can be accurately predicted.

Once the paper gage has been verified (therefore "calibrated") by a second independent check to catch any incidental errors, it can take on the status of any other functional gage.

If any number of parts meets essentially the same dimensional requirements, and are therefore acceptable by the same paper gage, the individual inspection reports can (1) reference the report that includes the paper gage or (2) include a copy of the original paper gage.

11.5.9 Paper Gages Compared to Other Functional Gages

Paper layout gaging offers advantages comparable to those of three-dimensional functional gaging and optical comparator gaging, without the disadvantages associated with the physical aspects of hard gaging.

The functional gage, however, has one distinct and unique advantage: Being a three-dimensional object, it receives a part exactly as a worst-case mating part would, thus giving direct evidence of intended function. The optical comparator and paper gage, as two-dimensional representations of the functional gage, are one step removed from a true functional gage.

Compensation for the value of the missing third dimension sometimes can be achieved on an optical comparator if the part is mounted so that the primary datum surface is perpendicular to the collimated light beam. The part must be moved so that an edge throughout the depth of the part can be examined in focus to ensure that it meets its locational requirement. Obviously, the part should be thin enough to be conveniently checked. The paper gage will include the three-dimensional effect of a functional gage as long as three-plane datum setup and axial perpendicularity information are included on the inspection report and in the layouts.

One disadvantage of the functional gage (in addition to the cost and lead time required to obtain one) is the necessity for gage tolerance and wear allowance, which cuts into its acceptance rates if a pessimistic gaging policy is applied. Neither the paper gage nor the optical comparator need include the tolerances or allowances, unless these types of gaging are being used as preliminary inspection devices prior to receiver gaging.

An important and unique advantage of the paper gage is that there is no significant time factor involved in making one; it can be developed quickly and easily for a single part with little planning and can be as quickly verified. Another advantage of the paper gage is flexibility of application. It is impractical to design and construct functional gages or chart gages for only one or two parts, but paper gages can be applied conveniently to 1, 100, or 10,000 parts. In the case of large numbers of similar parts, only a few paper gages may be required to check questionable parts. Finally, inspection report comparisons are more meaningful when paper gages are used.

11.6 SUMMARY

The techniques outlined in this chapter demonstrate inspection and measurement techniques that emulate hard functional gages. Each of the three methods contains both advantages and limitations of which the concurrent engineering team must be aware. The techniques provide additional tools available to the process planner, for both manufacturing and measurement that may overcome the economic limitations of hard functional gages. In making the appropriate choices for a specific project, these methods should be adapted in a manner that yields results similar to those obtained by true functional gaging. The level at which the alternative techniques simulate hard gaging is the deciding factor in which one of the methods to use and how it is to be applied.

REFERENCES

Tandler, W., Applying ASME Y14.5M to Coordinate Measuring Machine Operation, Menlo Park, CA: Multi Metrics, Inc., 2000.

12

Functional Workholding and Fixturing

12.1 INTRODUCTION

When following the normal sequence of events, production tooling is designed before any inspection equipment takes form. This book changes that order and intentionally presents the design of the functional gage prior to design of the tool. The gage design techniques presented in Chapter 9 focus on gage elements, to some extent neglecting design issues relating to the fixture. These examples demonstrate how gage design is driven by the product specification—in particular, information contained in the design layout. The underlying concepts are now extended to show how decisions the process planner and tool designer make are, in turn, driven by the gage design, particularly the fixture component of the gage.

The material emphasizes fundamental design principles usually associated with dedicated tooling. This is done while being fully cognizant of the tradeoff between the flexible tooling that supports state-of-the-art machining centers and the dedicated tooling needed for traditional machine tool designs. However, the fundamental principles can be seen in either type of setup equipment when good design and manufacturing practices are followed.

This chapter does not include extensive coverage of jig and fixture design. Many available texts go into the hardware-related details of designing production tooling (Hoffman, 1996), most with a perfunctory chapter on GD&T. These books do not demonstrate, however, how geometric control can be used as an integral element of the tooling phase of product development. To deal with this shortcoming, the following material emphasizes fixture design based on geometric control. The subsequent design concepts are a derivative of the ability of GD&T to tie process planning and tool design to the product definition.

The illustrated techniques can be applied to tooling for fabrication, assembly, and gaging. In fact, after designing the functional gage—the means to both locate and gage the part—much of the work involved in tool design is already done. These two apparently distinct fixtures are of a single design origin, the part's mating component. The gage design, in a sense, limits the ingenuity of the tool designer to stray from the functional intent of the product designer. The positions of the gage and tool in the design sequence (Figure 12-1) are reversed to emphasize the impact of geometric control on the complete development process. The product designer's explicit statement of required tolerances (both type and level) becomes meaningful only when the gage is designed. The gage provides one of the most efficient checks for component producibility.

An additional point of emphasis, which may have already occurred to the reader, is that the DRF, which has been the focal point of all the design techniques, is in fact the fixture. Each reference frame contained in the product definition corresponds to a fixture that needs to be created in some fashion, either with an actual fixture or through methods such as surface-plate inspection, to produce and measure the part. In the perfect world, a single fixture design would serve all purposes. Although the manufacturing world is not perfect, this ideal is more achievable than is commonly supposed; also, as demonstrated in a subsequent case study (Chapter 13), sometimes the ideal is achievable. What follows puts a physical reality to this concept.

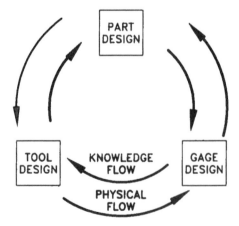

FIGURE 12-1   Design and implementation cycle.

12.2 FUNCTIONAL FIXTURES

A functional fixture can have one of two important purposes. As mentioned, it is one of the major elements of a functional gage. In this capacity it provides the physical means to locate (i.e., control translation and rotation) the part being inspected, putting it in the functional location it assumes in an actual assembly. The other application is to place either the rough stock or work-in-process in this same location—or in an equivalent location to be discussed later—for each of the operations in the manufacturing sequence, differing only to the extent that chip relief or other concessions to production reality must be made. In an ideal process, these two applications are satisfied by essentially the same design. Thus, the gage fixture and the tooling fixture would be identical, differentiated only by the level of control required for their execution.

The thread that wends its way throughout the entire design sequence is the geometric similarity of the part's mating component and the fixture embedded in the gage and the manufacturing tooling. Without this commonality of locating method, each designer involved in the production sequence may choose to introduce different methods of locating the part. These different techniques will cause errors in location that propagate throughout the manufacturing process. The result is a lack of certainty about the part's location in relation to its mating components (or processing equipment) and an ensuing stackup of tolerances. The worst situation involves so much variation that the parts will not assemble properly, resulting in rework and unnecessary expense.

It follows that controlling the location of the workpiece is the primary concern of any individual involved in the design of the manufacturing process (including inspection). If effective control is provided, then the least amount of variation will be exhibited within the manufacturing sequence. Tolerance stackups are minimized, reducing problems associated with both the manufacturing process and subsequent inspection.

The overarching theme of this chapter continues to be the control the design layout exerts over all the detailed design work leading to an acceptable product. Containing both the component to be tooled and its mating part, the layout describes all the geometric elements necessary to detail the part and to design functional gaging and tooling. This is an extremely important—but commonly overlooked—point; it is also why much of the product's cost structure is unknowingly locked in early in the development process.

12.3 FUNCTIONAL FIXTURING PRINCIPLES

In parallel with the functional gaging principles listed in Chapter 9, a set of equivalent principles underlies the design of what may be called "functional fixturing." It can be seen from the following list that many of the fixturing principles are the same as those for gage design. In most instances, inserting the word "fixture" in place of the word "gage" is sufficient to create the comparable fixturing principle.

1. Gages, production tooling, and parts (all of which may include tolerances and wear allowances) should be designed simultaneously using a concurrent engineering team.
2. The tool designer should not have to make arbitrary decisions regarding fixture element size, geometric characteristics, or location. A complete product specification dictates the fixture criteria through the use of appropriate geometric controls.
3. Fixtures should be defined using the same geometric characteristics used on the part being fixtured and its corresponding gage.
4. Functional fixtures have companion features (with respect to the part) that provide the datum feature simulators incorporated within the fixture. These datum feature simulators represent features on the mating component.
5. Functional fixtures have fixed positioning elements located at basic dimensions and conforming to feature locations described in the product definition.
6. These fixtures simulate location of the worst-case (virtual condition) part if there is no fit allowance or the worst-case (virtual condition) mating part if there is a fit allowance. The fixture is one of the elements comprising the design of the part's corresponding functional gage.
7. One datum reference frame per part will enable one fixture to be used for manufacturing. Any increase in the number of datum reference frames will increase the number of fixtures and setups (manufacturing, measurement, and gaging).
8. All functional fixturing elements that provide part location should incorporate the correct feature relationships through the use of the appropriate datum reference frames (both datum feature simulators and their precedence) or equivalent manufacturing techniques that serve to reduce setup error.
9. A conscious decision should be made to establish tooling tolerances and wear allowances that reflect the chosen gaging policy. This decision may be set by contract or by reference to appropriate standards.
10. Parts that can be practically tooled can also be practically gaged because fixtures and gages should be "interchangeable."

12.4 FIXTURE DESIGN CONCEPTS

The remainder of this chapter focuses on the geometric techniques that should accompany good tool design practice. In concert with established themes, the critical elements in the fixture design process are the identity of the functional DRF and the processing sequence. This functional frame is also the keystone of the design layout and the component design methodology. As a consequence, critical information necessary for functional fixture design already exists as part of the development cycle's knowledge base. The concurrent engineering team has already defined elements of the fixture as byproducts of earlier design stages in the development process. The following discussion extends and expands details developed in earlier chapters.

The degrees-of-freedom approach is used to establish the desired kinematic location of the part. Chapter 3 describes this in general terms and presents the foundation ideas. To flesh out these concepts, the example in Figure 12-2 will be used to illustrate details not covered earlier. Previous examples deal with some common geometric shapes unadorned with additional features. It is pointed out earlier that these shapes easily lend themselves to the 3-2-1 system of positioning described in most texts on jig and fixture design. Unfortunately, many of the most critical design problems involve parts that do not fit into such convenient categories. The increased complexity of part geometry in the more difficult design problems requires greater understanding of the theory underlying the 3-2-1 system (Eary and Johnson, 1962).

The first example is a common shape, a prismatic solid (Figure 12-2), which uses a combination of a functional feature, a tooling hole, and a tooling slot to establish the reference frame. Tooling holes are used as secondary and tertiary datums on most aircraft parts and over 50% of all automotive parts. Unfortunately, in many cases design and quality engineers are not aware of the need for tooling holes. As a consequence, the manufacturing engineer determines which

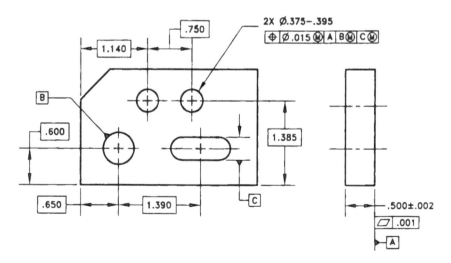

FIGURE 12-2 Sample part to illustrate degrees-of-freedom tooling concept.

holes to use and may default to existing holes that are farthest apart or may even add holes if suitable ones are not available. It would not be unusual for this tooling information never to get back to the design engineering staff and never be documented in the definition of the finished product. Strange things can happen operationally when matching holes are chosen in mating parts not for enforcing function but rather for manufacturing convenience. One of the authors encountered a situation where the holes chosen for tooling applications caused a mismatch between the heads and the engine block of an automotive engine. The engines failed after several thousand miles, leaving customers less than pleased with the company's product.

Look at Figure 12-2; the design goal is to control or eliminate the six degrees of freedom (three translational and three rotational) present for any manufactured component. In Figure 12-3 the part has been placed on a tooling plate that simulates the primary datum. At this point, the base element of the fixture—the tooling plate—will prevent motion along one direction of the z-axis; the part cannot move into the tooling plate. Additionally, if the part remains in contact with the plate, there can be no rotation about the x- and y-axes. Control has been established (Figure 12-4) over 2 1/2 of the 6 degrees of freedom; no rotational motions are allowed about the x- and y-axes, and no motion is allowed in one of the two directional senses associated with translation along the z-axis.

If a fixed pin were now inserted into the plate to pick up the tooling hole, additional motions are constrained. The part can no longer translate along the x- or y-axis, eliminating two more degrees of freedom (Figure 12-5). The final constraint the fixture must provide comes from a second fixed pin (e.g., a diamond

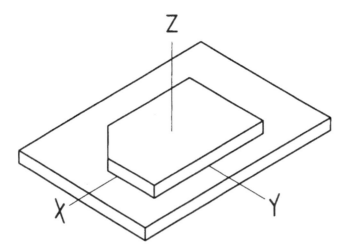

FIGURE 12-3   Part placed on tooling plate.

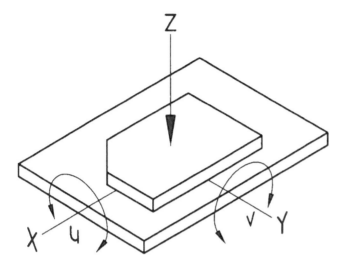

FIGURE 12-4 Degrees of freedom controlled by tooling plate.

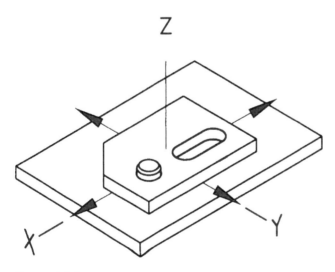

FIGURE 12-5 Degrees of freedom controlled by tooling hole.

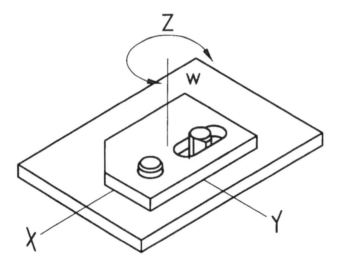

Figure 12-6 Degrees of freedom controlled by tooling slot.

pin) that contacts the slot. The pin is oriented so that it can contact both opposed planes that establish the slot's width dimension. The pin prevents rotation about the z-axis (Figure 12-6), bringing control to 5 1/2 of the 6 degrees of freedom.

The last possible motion (Figure 12-7) is along the z-axis. This motion is necessary to load the part on the fixture. How this z-axis motion is controlled depends on the use of the fixture. If the fixture is used in a machining process, a holding force introduced by a clamp may be provided. If the tool is an inspection fixture, the weight of the part may prevent motion in this direction and the fixture will require no additional components.

For the sake of completeness, it should be mentioned that some tooling texts identify 12 degrees of freedom, separately counting each of the directional senses; there are two possible motions along the x-axis (plus x and minus x), and so forth. Under this set of circumstances (Figure 12-8), the tooling plate removes five degrees of freedom, the circular pin removes four, and the diamond pin takes two. Thus, 11 degrees of freedom are controlled, leaving only the motion in the +z-axis to be dealt with.

Another example, more geometrically complex, is shown in Figure 12-9. This component is produced by a progressive die and retained on the strip. Subsequent assembly operations work from a reel containing the components. The fixture is the die, with the datum feature simulators consisting of the die blocks in the lower shoe (primary datum feature simulator) and the pilots (secondary datum feature simulators) controlling the location of the strip. A stripper would be equivalent to a clamp to hold the strip against any manufacturing forces. This

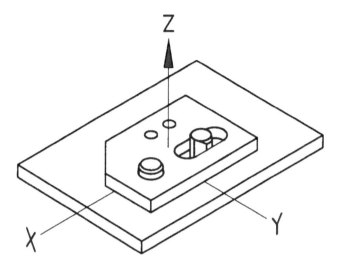

FIGURE 12-7   Last degree of freedom controlled by clamp.

is a good example of a single-process DRF used to control the simultaneous manufacture of the functional DRF and the features related to the functional frame. Control of location is achieved in a fashion similar to the previous example. In this instance, the pilots perform the combined function of the tooling hole and the slot. Because all the pilots are of the same size, it is not possible to differentiate between the secondary and tertiary datum features; each pilot is of equal precedence.

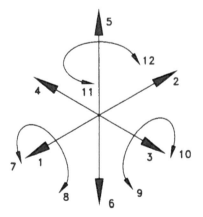

FIGURE 12-8   Tool designer's approach using 12 degrees of freedom.

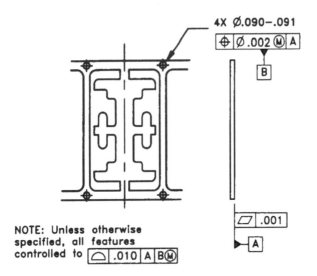

FIGURE 12-9   Example of process DRF taking precedence over functional DRF.

12.5   DESIGN DETAILS

The type of DRF to be used in designing the fixture is intimately tied to the concurrent design process. The advantages, where feasible, of using the functional DRF throughout the product development cycle are obvious. When a fixture is created, one of the initial issues to deal with is the choice of the DRF on which to base the design. This is actually two separate decisions: The first deals with the type of DRF to be used; the second deals with the number of DRFs to incorporate in the manufacturing process. Both of these decisions are either dictated or heavily influenced by the product definition created in the early stages of the development process.

12.5.1   Functional Versus Process Frame

The philosophy of this book is to use the functional DRF in the design of all gages and fixtures. However, there are times when the processing sequence makes the choice more ambiguous, possibly eliminating the preferred functional frame from consideration. In particular, the design may require team consideration of process DRFs, rather than the functional frame, to achieve the dimensional control specified in the product definition. An example of process requirements overriding the preferred functional DRF is the example in Figure 12-9, where the component is retained on the strip and wound on a reel. The reel is then incorporated into the design of an automated assembly machine that uses the strip's process DRF to provide location in the assembly operation.

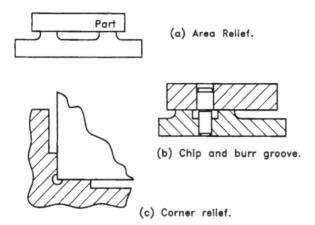

(a) Area Relief.

(b) Chip and burr groove.

(c) Corner relief.

FIGURE 12-10   Examples of chip relief affecting DRF.

Another situation might exchange the functional DRF for better geometry control. An example might be a part where the primary process is casting or forging. The resultant surface characteristics of a casting may require placing locators (datum feature simulators) as far apart as possible to minimize the effect (i.e., sine or cosine errors) of surface irregularities on the dimensional specifications. The design impact of the irregularities manifests itself by the presence of datum targets in the detailed component definition. A further example involving a process DRF differing only in degree from the functional frame is where chip relief is incorporated into the fixture design. This relief might be obtained by using machined pads (actually datum targets) to locate the part rather than a continuous planar surface (Figure 12-10) that simulates the surface of the mating part. In the latter case, the machined pads should be specifically identified on the product drawings as targets.

In many cases features considered necessary for good tool design practice (e.g., relieved locators for chip relief) are treated as standardized elements and not explicitly incorporated into the finished product specification. As a consequence, the tool designer introduces subtle changes in location of which the design team may be unaware. Such examples of accepted tool design practice, while possibly necessary, should be explicitly incorporated into the design. This enforces documentation of the design feature and allows for subsequent analysis of the technique's effect on overall product variation.

12.5.2   Number of DRFs

After concluding the discussion involving functional versus process DRFs, the team should consider the number of DRFs to be incorporated in the manufacturing process. This decision is made in parallel with decisions relating to the manufacturing sequence and the type of DRF. The choices are (1) a single, functional DRF that maintains the design integrity among the part, the functional gage, and the functional tooling, (2) a process DRF that breaks the link between the gage and the tooling, and (3) multiple DRFs that could be a combination of both functional and process datums. Without a doubt, the preferred method is the single, functional DRF. When the functional frame is not easily incorporated into the process design, machine tools possessing multiple processing capability (e.g., turning and milling) in a single machine may provide an attractive alternative. Using the single-setup capability of such machine tools, a single, process DRF might be used which would allow all the features that comprise the functional DRF and all the features related to the functional DRF to be generated in one setup. Thus, the desired functional relationships are dependent on machine capability—no process tolerance stacks are introduced. This technique is explored in Sec. 12.7.2 and in Chapter 13 as part of a case study. Both of these expositions support the alternative technique alluded to in item 8 of the fixturing principles.

A tool designer's choice of multiple setups (tooling and gaging) for a specific component necessitates more complicated analysis methods to predict the outcome of the production process. Multiple setups introduce additional real, not just hypothetical, variation into the process. It has been the authors' experience that the greater the number of DRFs used in both defining and producing a part, the more likely the design team is to resort to an "analysis by experiment." The components are made without prior tolerance analysis and then assembly is attempted. If the parts do not fit, the necessary changes to the product drawings and hard tooling are made. Another prototype run is performed, and the process is repeated until satisfactory assembly is achieved. The resort to additional DRFs is a rather insidious way to avoid the thought inherent in a well-developed product design.

The result of the decisions concerning the reference frame, when based on analysis, will ultimately be a fixture that

1. Provides kinematic control of the part
2. Consistently places the part in its "next assembly position"
3. Can be used to control the location of the tool (e.g., milling cutter) or inspection instrument relative to the part

The last item is of particular concern because considerable variation can still be present as a result of the characteristics of the production process. The fixture presents the part in the correct location for processing but cannot counteract the effects of ill-chosen or inappropriate processing sequence or methods. It is only one of many elements in the processing system that must be carefully selected to ensure successful production.

12.5.3 Location of the Fixture

Another design issue concerns the process DRFs inherent in any fixture design. The reference frame controlling workpiece location is itself located within a reference frame established by the fixture body and its elements. In turn, the fixture is located with respect to the processing equipment. This series of process DRFs must also be consciously designed and evaluated for its effects on process variation.

The most visible elements of the fixture's reference frame might be features such as setblocks, keys, and tooling balls (Figure 12-11) that allow the fixture to be placed in the correct relationship within the manufacturing system. These tooling elements establish the fixture DRF and allow the fixture to be placed within the machine tool/fixture assembly. In more traditional jig and fixture design, such features may be treated as standard design elements—as mentioned previously—and enter the design cycle without sufficient consideration and analysis.

Less visible elements of the design include the desired form and geometric relationships the datum features and datum feature simulators comprising the fixture must establish. Some of the tenets of good design practice may require specifying the necessary functional relationships of the fixture and the specific

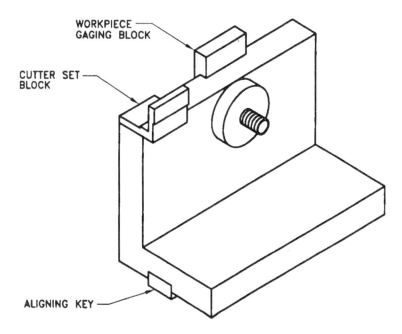

FIGURE 12-11 Fixture demonstrating set block, key, and gaging block.

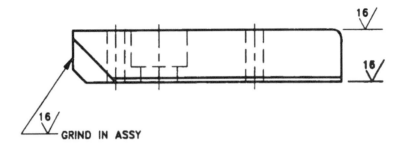

FIGURE 12-12   Partial detail of set block requiring grinding in assembly.

manufacturing techniques to impose these relationships on the tooling. This is demonstrated in Figure 12-12 by the callout requiring grinding after assembly to achieve the appropriate precision level and to ensure the necessary geometric control among the fixture components.

12.5.4   Design of Datum Feature Simulators

Fixture elements that locate the part (datum feature simulators) should mimic the full extent of the contacting feature found on the mating component. Where a functional DRF is used, the fixture will take on the appearance of the mating component at the part–mate interface. These datum feature simulators must be accessible and of sufficient size to serve as reliable features. This is a reasonable statement—as it comes from material contained in the Y14.5M standard—and common sense would dictate that a designer should adhere to such rules when choosing datums.

A story serves to emphasize the importance of these datum issues. One of the authors was asked to review drawings of a product that was the subject of a lawsuit. The defendant was the company that produced a gage necessary for functional inspection. In looking at the product drawings, it appeared that every one of the 30 or so rational combinations of the 13 available geometrical toler-ances (the 1982 standard) had been used on each of the part drawings forming the assembly. Further examination found that the DRF contained in the product definition was inadequate, bringing a type of circular logic into the datum speci-fications. The worst error was a tertiary datum buried in the forward portion of the major component and inaccessible for gaging purposes. The lawsuit was brought against the gage manufacturer because the gage did not locate the tertiary datum. The gage manufacturer made a phone call to point out the access problem and request drawing changes. However, engineers at the contracting firm remained unconvinced of the problem and no resolution was reached. Ultimately, the gage company lost the case and is now out of business.

In choosing datum feature simulators, it is important to remember that the DRF establishes the coordinate system used to take measurements. The resulting measurements are as critical to part location in manufacturing as inspection. Once the part is placed within the coordinate system (the DRF), subsequent measurements do not intrinsically account for uncertainty the datum features introduce. In a limited sense, the datum features are assumed to be perfect for the purpose of either measuring the part or achieving the desired locational relationship between the part and the machine tool. Hence, the appropriate choice of the datum feature simulator is of paramount importance in limiting the effects of process variation.

The accuracy level for fixture elements is dictated by the product definition and depends on whether the fixture is used in gaging or tooling. It would not be unusual to see a 5 to 1 or 4 to 1 accuracy ratio used in the production tooling, with a 10 to 1 ratio required by gaging policy. The last ratio might be reduced if the tolerances are so precise as to impose unreasonable requirements on measurement capability or the traceability chain.

When a casting or forging is used as rough stock, the process plan may require qualifying selected surfaces for use as datum features. However, if the full extent of these machined features is not used in the subsequent assembly, nonfunctional cost is added to the product. Datum targets (which must be explicitly described in the product definition) are used to overcome this criticism by providing a reliable, repeatable, and less expensive method of locating the part. Note that an analogous location technique is also used in general fixture design (Figure 12-10) to reduce the surface area on which chips or dirt could accumulate and cause errors in setup.

Many of the small design details (e.g., chip relief and relieved locators) typically present in tooling designs raise questions about the quality of the datum feature simulators. These modifications to a simulator must still reflect the mating features found in the design layout and should be of appropriate form and extent to provide the required component location. Any alteration of datum feature simulators for manufacturing purposes should be done only by consensus of the engineering team. Design features based on standardized practices of the tool design department require clear discussion of the impact of such changes on component variation, and their use should be subsequently documented in the product definition.

Methods of indicating datums on an engineering drawing are tied to explicit definitions of the extent of the datum. An example is shown in Figure 12-13(a), where the lower surface of a bracket is shown as the primary datum. The full extent of this surface, because of the drawing specification, corresponds to the datum that locates each part as it is assembled. For such an application, the quality of the datum feature might require form control to meet functional requirements. This geometric control would bound deformation in the area of the bend. If the

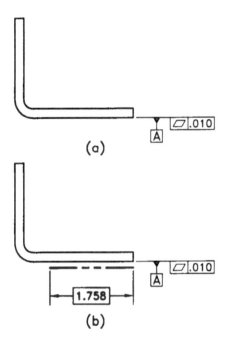

FIGURE 12-13   Specifying extent of datum feature.

drawing specification is changed because the part does not rest on the entire datum feature, controlling form variation over the entire surface adds cost but has no functional effect; assembly contact is not made in the area of the bend. This change in design intent can be communicated by the use of a chain line indicating that only a portion of the surface acts as the functional datum [Figure 12-13(b)]. The change in specification may entirely eliminate the expense of a secondary operation to achieve control of flatness.

Another case is where a component mounts to its mating part and contact is made with the entire area of the primary datum surface. In the event that subsequent operations use this surface as part of the DRF, the tool designer must consider the effects that result from using area locators (targets) to eliminate chip and dirt problems (Figure 12-10); the location of the part is now determined by a smaller point set created by the datum targets. The primary datum would be established only by those points contained in the target areas rather than the full extent of the datum feature. In such a situation, the fixture design would not comply with the product definition if targets were not specifically identified in the specification. Even with the targets included in the product definition, there will be a difference between the desired functional location and the location provided by the process tooling. The result is additional uncertainty associated with the part's location. Depending on the level of control the product definition requires, this may not be of concern. However, it is an issue that must be consciously considered rather than being decided by default.

12.6 APPLICATION ISSUES

Elements comprising a fixture perform one of three functions: They locate the workpiece, support it, or clamp it, preferably without distortion. Most of the nut-and-bolt issues are more than adequately covered in books dealing with jig and fixture design. This last section is intended only to touch on some of the points that reflect issues raised throughout this book.

Control of the elements that make up the fixture and locate it in relation to the machine tool are as important as control of the workpiece by the fixture. Within the fixture DRF, each fixture component should be accurately located and exhibit the required control (size, form, orientation, and location) necessary to perform the workholding functions. Each fixture component is effectively assembled within a system comprised of the fixture and the machine tool. Keys or other means should be provided to accurately position the fixture on the machine tool and setblocks or tooling location points provided to create the necessary relationship between the fixture and the process tools (e.g., milling cutter, EDM electrode, etc.). The result of the complete fixture design—including the machine tool—is a set of chained DRFs requiring explicit interrelationships to produce acceptable product.

The fixture should incorporate the means to limit or control variation that may be unique to a specific manufacturing process. Examples would include parting lines and flash that inhibit reproducible location of the part in the fixture and that are found in cast, forged, or molded components. The mislocation caused by these characteristic process features prevents the part from being placed in the singular functional DRF called out by the product definition. Obvious methods to counteract these locational effects include using datum feature simulators that either mitigate or remove the effects of the unacceptable process feature. Datum features that do not include the unwanted process feature or datum targets provide reproducible location by carefully defining the product.

The engineering team should consider running a force/deformation analysis. This is done to assess impact of tool forces, gravity, thermal loads, and other effects inherent in the part/processing system design that may statically or dynamically alter feature location and relationships during processing. In processes that generate significant force loading, the fixture should be designed to direct both tooling and clamping forces into the locators and away from the clamping devices.

In the event a part exhibits significant deformation, making it difficult to conform to the product definition, consider the tradeoff between part quality and production costs by introducing support or restraint. The supports may be either fixed or adjustable. Supports must be incorporated in a manner that prevents them from becoming alternate locators that may vary the DRF from setup to setup. Any restraint imposed on the component for processing should be documented in the product definition.

Where the fixture is designed to stack or gang parts together, locate all workpieces by positive and independent means when designing processing fixtures. Each of the workpieces must be located by an independent DRF.

Make sure (1) that in any design, part contact with the locators can be verified by the operator and (2) that this contact is along the full extent of the datum feature as described by the product definition. The operator must know if the part is properly located within the DRF.

The fixture design should allow for measurements to be taken in the fixture for both process control and product conformance purposes.

12.7   A PRACTICAL EXAMPLE

A single DRF used in manufacture and inspection is obviously a goal that cannot be imposed in all situations. To illustrate how the functional fixturing concepts can be applied to more commonly encountered parts, the details shown in Figure 12-14(a) highlight the general usefulness of the chapter's techniques in designing tools or gages. The discussion first approaches part processing from the traditional standpoint, using concepts from this chapter to explain what each operation does in terms of establishing the DRF and subsequent control of the functional features. The process steps are predicated on manufacturing technology as it existed circa 1970, the time period when the part was first designed and processed.

There is a method to this approach, even though analyzing an outdated process plan would seem to contravene the ideas contained in the text. The first element of the rationale is that each of the steps chosen for the more traditional, manual equipment process accomplishes a specific purpose relative to the functional requirements of the design, but was done within the context of the facility where the part was produced. This context encompassed the firm's available manufacturing environment—manufacturing personnel, machines, facilities, etc.— and the prevailing economics may not have allowed the option of buying new equipment or subcontracting. As a result, the process planner would have needed to display an ingenuity that added more steps to the process plan, accounting for specific capabilities and limitations within the plant. Each of these operations contributed, albeit by introducing additional setups, to part functionality.

Another reason for the discussion, even when applying current technology, is that a planner may be constrained to use existing equipment where functional fixtures may still be necessary. The process planner and tooling engineers must understand and be able to articulate the relationships contained within the design. They have to consider the impact of these relationships on process planning. This

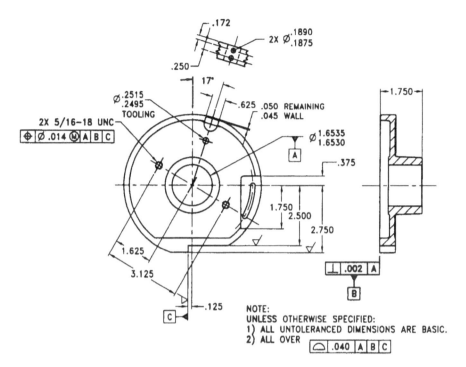

FIGURE **12-14(a)** Partial detail of example part.

requires a detailed knowledge of the sequence within which the DRFs are generated and of how the process controls the functional features relative to the various DRFs.

There is no disputing the fact that the traditional process contains steps that even when measured against the state-of-the-art in the 1970s might not be required given newer or better-maintained equipment—a more optimum manufacturing environment. To criticize the example is to miss the point. Before the process can be changed to take advantage of newer capabilities, the purpose of each operational step must be understood and described in a language that clearly identifies the desired outcome vis-à-vis the part definition. Newer equipment is not going to eliminate critical elements of the design or the need for the manufacturing engineer and the process planner to understand the intricacies of the part geometry. This part geometry includes both the permanent features incorporated into the part definition along with transient features necessary to carry out the process plan. To develop this level of understanding, the following example is processed using both a general and now-superceded approach and then using equipment that allows for single-setup manufacture.

12.7.1 Traditional Processing of Part

The primary process for this detail is aluminum casting [Figure 12-14(b)], which creates a part 6 in. in diameter. The first process plan to be analyzed is described in Figure 12-15. The initial machining operation takes place in a lathe with work-holding provided by a three-jaw chuck. The chuck establishes a process DRF consisting of the external diameter of the cast hub (the primary datum axis) and the planar feature that forms the backside of the drum—the secondary datum feature, which stops motion along the datum axis. No tertiary datum is required because no asymmetric features are affected by this setup.

The first machining takes place in Operation 20 and is shown graphically in Figure 12-16(a). The outer face of the rim and the outer face of the hub (i.e., both to the right in the figure) are faced off during a roughing operation. The rim is also rough-turned using the same process DRF. The inside diameter of the hub is then rough and finished-bored and reamed to size while still in the same setup.

From a purely geometric standpoint, the first operation uses a process DRF to create the primary datum feature of the functional DRF. Subsequent operations continue to use this critical feature (i.e., the hub bore) to provide the necessary dimensional and geometric control.

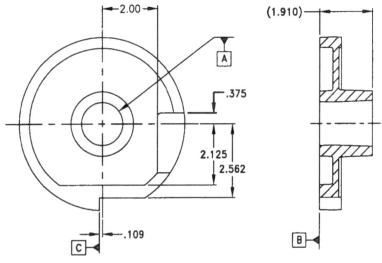

NOTE:
UNLESS OTHERWISE SPECIFIED:
1) ALL UNTOLERANCED DIMENSIONS ARE BASIC.
2) ALL OVER ⌲ .060 A B C

FIGURE 12-14(b) Partial detail of example forging. Detail of casting.

ABRIDGED PROCESS SHEET		PART NO.	Figure 12-15	
DEPT. NO.	OPN NO.	OPERATION		
	10	Requisition from stores		
	20	Face off outer face of rim; Face off outer face of hub; Rough & finish bore; Ream bore; Rough turn O.D.		
	30	Face off inner face of hub; Face off inner face of rim (stock allowance on both operations for finishing)		
	40	Face off outer rim to finish; Face off outer hub (stock allowance for finishing); Turn O.D. to finish dim		
	50	Face off inner side of hub to finish		
	60	Spot Drill, Drill, Ream tooling hole; Drill two letter "F" holes		
	70	Tap 5/16 - 18 holes		
	80	Mill step into O.D.		
	90	Mill flat on O.D.		
	100	Remove burrs		
	110	Mill Step for .187 x 30° Slot		
	120	Mill .187 x 30° Slot		
	130	Remove burrs from slot		
	140	Inspect		
	150	Deliver to Stores		

FIGURE 12-15   Abridged process plan for workpiece in Figure 12-14(a).

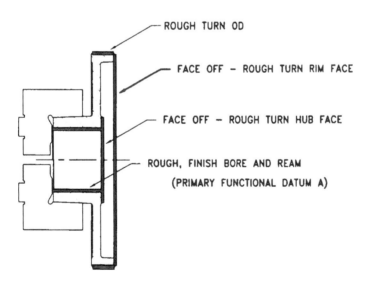

FIGURE 12-16(a)   Drum example, Operation 20.

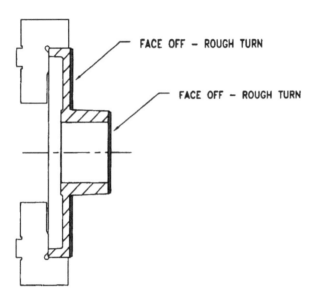

FACE OFF – ROUGH TURN

FACE OFF – ROUGH TURN

FIGURE 12-16(b)   Drum example, Operation 30.

The inner faces of the rim and the hub are rough-turned by reversing the part's position [see Figure 12-16(b)] in the machine tool. During this operation (Operation 30), a three-jaw chuck contacts the machined rim using previously prepared soft jaws. This creates a new process DRF established by the rim—the rim qualified but not finished-turned—and the outer face of the rim, both previously machined in Operation 20.

Operation 40 [Figure 12-16(c)] returns the part to the initial position of Operation 20 using an expanding mandrel in conjunction with the reamed hub bore to simulate the primary datum of the functional DRF. The facing operations that take place at this point in the sequence include work on the outer face of the rim and the outer face of the hub. Note that the outer faces are finished-machined, although a stock allowance is left in the finish machining dimensions; this makes the required dimensional control possible when subsequently machining the inner faces. The 5.705-in. outside diameter of the rim is machined to the finished size.

At this stage of the process, two of the features comprising the functional DRF (the bore and the outer hub face) have been created; the only exception is that no feature (the tertiary datum) has been provided for clocking of the component.

The next operation [Operation 50; Figure 12-16(d)] reverses the position of the drum, again locating on an expanding mandrel. The inner faces are finished-

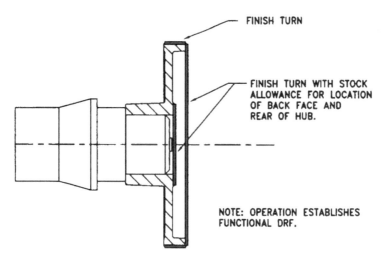

FIGURE **12-16(c)**   Drum example, Operation 40.

turned, providing the necessary dimensional control relating these features to the functional DRF. None of these features involves a specification of angular position; as a consequence, the absence of the functional, tertiary datum feature has no effect on control of the features generated in this setup.

To create a reference frame equivalent to the functional DRF, Operation 60 [Figure 12-16(e)] spot-drills, drills, and reams a ⌀.2505-in. tooling hole. If

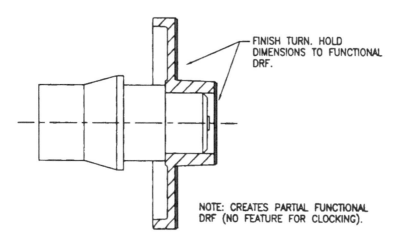

FIGURE **12-16(d)**   Drum example, Operation 50.

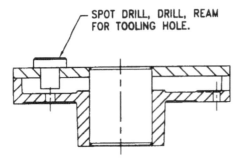

FIGURE 12-16(e) Drum example, Operation 60.

a drill jig were to be used to control the accuracy of this operation, its design (Figure 12-17) would incorporate the features of the functional DRF elements previously machined. The tooling hole serves to create the equivalent of the tertiary datum feature needed to complete the functional frame, although the hole is not the intended functional feature. At this point, subsequent operations can make use of this reference frame—equivalent to the functional DRF—to ensure the desired dimensional and geometric control.

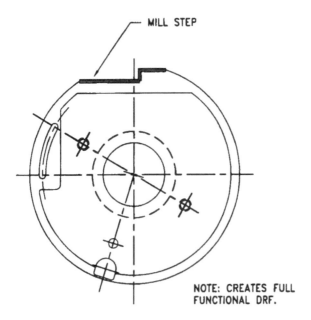

FIGURE 12-16(f) Drum example, Operation 90.

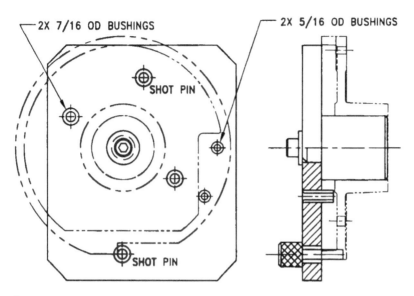

2X 7/16 OD BUSHINGS

2X 5/16 OD BUSHINGS

SHOT PIN

SHOT PIN

FIGURE 12-17   Drill jig for drum.

Operation 70 is not illustrated here. However, the tapping operation requires another setup. It does not need another fixture for location purposes but does introduce a time penalty that adds to total processing time and cost.

The complete functional DRF is finished in Operations 80 and 90 [Figure 12-16(f). A step is milled on the rim of the part, establishing the last feature needed to complete the functional DRF. In this setup, the bore, the outer hub face, and the tooling hole serve as the process DRF—although containing two of the three features in the functional DRF—used in the design of the workholding fixture (see Figure 12-18 for the milling fixture). The complete functional reference frame is now available for subsequent operations and can be used to create fixtures that conform to both the part definition and good design practice.

The final functional features (Operations 110 and 120) will be placed on the workpiece using another milling fixture that can utilize either the functional DRF or the same process DRF used in Operation 90. In the latter case, the fixture would contain the same elements as the milling fixture illustrated in Figure 12-18 but would be configured for a vertical mill. This fixture could be mounted on a rotary table to allow the same fixture used to mill the step for the slot to also perform the fixturing function for Operation 120 where the circular slot is milled.

While it may not be obvious, this example continues to illustrate use of the functional DRF in fixture design. Even in situations where the single-frame concept cannot be implemented directly, it provides a focal point to use in the

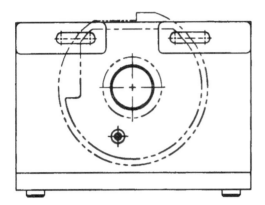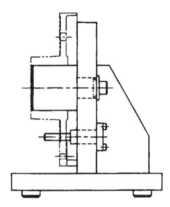

FIGURE 12-18   Milling fixture for drum.

design process. It creates a benchmark that guides analysis and decision making during design of both the process plan and supporting tooling. The goals of the various analyses—of the process plan, resulting manufacturing sequence, and tooling requirements—are to minimize the effects of tolerance stackups and limit the total amount of variation exhibited in the manufactured components.

12.7.2   Single-Setup Processing of Part*

As mentioned, existing equipment may be used to process a part because of business limitations. This will inevitably result in compromises that test the creativity of the process planner. However, if economics allow, there are alternative technologies that can increase process efficiency along with improving the accuracy with which the geometry can be created. One such alternative might be the use of a multispindle, multiaxis lathe capable of machining the part in what could be termed a single setup. An example of this type of machine tool is shown in Figure 12-19.

The resulting process plan (Figure 12-20) still contains a large number of process steps but no longer requires the multitude of setups found in the first processing example. Using the capabilities of a multispindle machine, it is possible to complete a large number of the turning, drilling, and milling operations in one spindle and, without releasing the part, pass it to the other spindle for completion. Without trying to optimize the process, one possible sequence would

---

* The authors would like to thank Randy Harland of Machinery Systems, Inc., for his technical assistance in assembling this section of the text.

FIGURE 12-19 Multispindle, multiturret turning machine. (Photo courtesy Mazak Corporation.)

ABRIDGED PROCESS SHEET		PART NO.	Figure 12-20
DEPT. NO.	OPN NO.	OPERATION	
	10	Requisition from stores	
	20	Face off outer face of rim (rough and finish); Face off outer face of hub (rough and finish); Rough and finish turn O.D.	
	30	Mill step for .187 x 30° slot	
	40	Mill .187 x 30° slot	
	50	Drill two letter 'F' holes	
	60	Tap 5/16 – 18 holes	
	70	Transfer between spindles	
	80	Face off inner face of hub (rough and finish), Face off inner face of rim (rough and finish);	
	90	Mill step into O D	
	100	Mill flat on O D.	
	110	Rough and finish bore I D , Ream I D	
	120	Inspect	
	130	Deliver to stores	

FIGURE 12-20 Abridged process plan for workpiece in Figure 12-14(a) using single-setup processing.

perform operations up to Operation 70 on one spindle and then have the part transferred to the other spindle for completion. At no time is the part permitted to seek a new orientation or location since the transfer is performed without allowing the part freedom to move in an unconstrained fashion.

The operations performed on each spindle are illustrated in Figures 12-21 and 12-22. In operations up to Operation 70, the part is oriented and located within a process DRF (shown in Figure 12-21) comprised of the face of the hub and the hub O.D. Rough and finish machining operations are performed by utilizing x- and y-axes, allowing both turning and milling operations to be performed in a single setup.

Once this set of operational steps is performed, the workpiece is transferred to the other spindle. The remainder of the turning and milling operations is completed using a different process DRF. While the part is not completely machined with respect to a single functional reference frame, the manner in which the second process DRF is created minimizes the uncertainty in part location to an order of magnitude that is not possible with the many setups involved in the traditional process. This second process DRF is established by the finished surface of the rim O.D. and the rim outer face (see Figure 12-22). It should again be emphasized that this DRF was acquired without removing the part from the setup. The part was transferred to the second spindle while being held by the first spindle. The chuck on the second spindle closes, and only then is the part released from the first spindle. While using two process DRFs, the movement of the workpiece

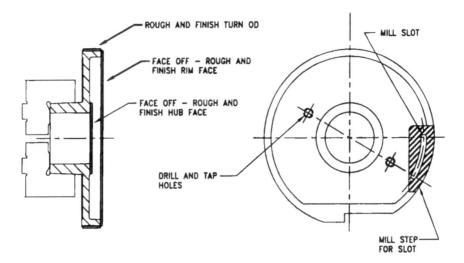

FIGURE **12-21**   First spindle; Operations 20 through 60.

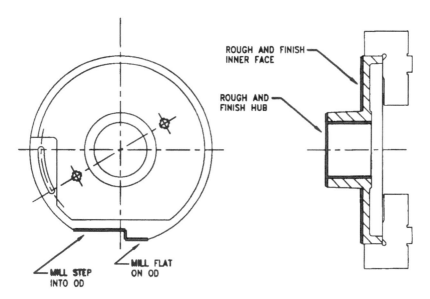

FIGURE 12-22 Second spindle; Operations 80 through 110.

within the machine is done in a manner designed to minimize orientation and positional errors.

The operations performed in the second spindle place the functional DRF on the part. This seems somewhat backward from conventional process planning where it is normally desired to acquire the functional DRF as early in the process as possible and then use it to generate the remaining functional features. However, the reduction in setups and the improvement in accuracy of the machine tool make it possible to perform the operations in this order and still meet the product specifications. In fact, the reduction in setups reduces the uncertainty of part orientation and location sufficiently to create a marked improvement in the control of the finished product.

12.8 SUMMARY

In considering the design of a fixture, either for processing, gaging, or measurement purposes, the critical consideration is the choice of a reference frame that will duplicate the effects of the functional DRF. The issues and concerns addressed by the concurrent engineering team relate to the appropriate design of datum feature simulators comprising this DRF and serving to reduce setup error in processing. It is up to the process planner and the tool designer to bring these issues to successful resolution, providing a physical reality to the product definition through the design of an acceptable process plan and the appropriate fixtures.

REFERENCES

Eary, D. F. and Johnson, G. E., Process Engineering for Manufacturing, Englewood Cliffs, NJ: Prentice-Hall, 1962.
Hoffman, E. G., Jig and Fixture Design, 4th ed., New York: Delmar, 1996.

13

Does It All Work?

13.1 INTRODUCTION

With all the ideas and associated detail that have been suggested, the question as to whether these concepts will really work in an industrial application is raised. The most effective way to illustrate the answer is with a case study. What follows is a demonstration of the methods as they were applied during a consulting assignment, showing how they are to be instituted and highlighting their rewards.

It is unlikely that any firm would make wholesale changes to an established product development system in an attempt to capture promised but unsubstantiated benefits. As a consequence, the most expedient way to foster the desired change is to apply this book's structured techniques to an existing product. Fortunately, these ideas apply as readily to the improvement of current products as to the design of new ones.

The most conservative approach to make the transition from the inefficient methods of the past to a more controlled future is to undertake improvement projects using both a simultaneous engineering team and the integrated design approach. This allows raising the confidence level of both managers and practitioners in circumstances where there will be less pressure to meet delivery deadlines at the risk of introducing design inefficiencies. It cannot be over-emphasized that what follows is the result of a concerted and cooperative effort. It is not a magic bullet that can be read about in the trade press on Friday and instituted on Monday. The results have to be earned through challenging, disciplined work.

The success any project achieves is obviously constrained by the choice of the project but also by the experience and knowledge of the participants. The following example is unusual in one respect. The outcomes of the project support all the contentions that have been made. It turned out to be the perfect example necessary to convince decision makers that this product development process works.

13.2 THE INITIAL SITUATION

Like most people, the managers of any company are hesitant to introduce massive change. After being exposed to the concepts associated with concurrent engineering, the managers at a specific manufacturing firm were reluctant to make changes to the way in which their products were designed and processed. Specifically, they were apprehensive about being able to estimate the cost savings attributable to these techniques. Because of the complicated nature of product development, these savings are not easy to gauge and such estimates are not without risk. So, taking a cautious approach, they asked to see the methods demonstrated.

The demonstration project focused on the assembly of the company's major product. Within the assembly, shown in simplified and abridged form in Figure

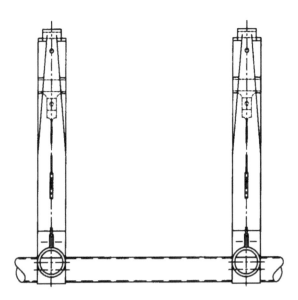

FIGURE 13-1 Simplified assembly.

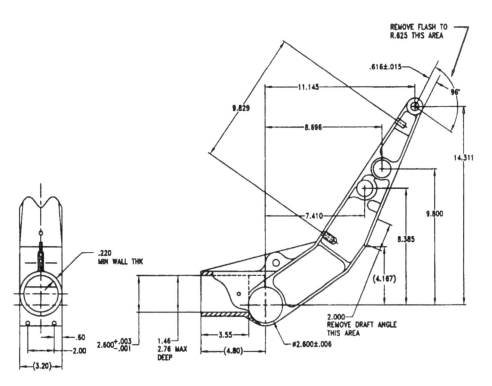

REMOVE FLASH TO
R.625 THIS AREA

.616±.015

96°

11.145

9.829

8.696

14.311

9.800

7.410

8.385

.220
MIN WALL THK

(4.167)

2.000
REMOVE DRAFT ANGLE
THIS AREA

.60

2.600⁺·⁰⁰³₋·⁰⁰¹

1.46

3.55

ø2.600±.006

2.00

2.76 MAX
DEEP

(4.80)

(3.20)

FIGURE **13-2** Abridged detail of workpiece showing selected dimensions.

13-1 to disguise the company's identity, the support was determined to be the critical part since the assembly is built off this base. The chosen component (Figure 13-2) actually turned out to be a family of parts that numbered approximately 56 distinct variations produced in the same manufacturing cell. The impetus behind this choice was failure to achieve the levels of productivity used in justifying the cell's purchase. The predicted gains had not materialized, and it was hoped that the project could correct the situation. Further support was gained when company personnel pointed out that the large product variety was causing manufacturing and logistical nightmares.

To highlight the scope of the problems with the cell, Figure 13-3 shows the abbreviated process plan that led to the cell design shown schematically in Figure 13-4. Each of the major pieces of capital equipment required a separate fixture. Note that the cell does not contain an inspection process, also requiring setups for each of the characteristics identified on the engineering drawing. The individual fixtures, both tooling and gaging, introduce new DRFs that place the part in different locations in space. Each reference frame must be analyzed for

ABRIDGED PROCESS SHEET		PART NO.	Figure 13-3
DEPT. NO.	OPN NO.	OPERATION	
	10	Requisition from stores	
	20	Drill	
	30	Broach	
	40	Rough Drill	
	50	Drill	
	60	Bore	
	70	Radius	
	80	Mill	
	90	Rod Press – insert pivot pin	
	100	Broach	
	110	Drill and Tap	
	120	Mill	
	130	Drill – point to point	
	140	Mill	
	150	Degrease	
	160	Grind	
	170	Peen	
	180	Clean, Package, move to stores	

FIGURE 13-3   Abridged process plan listing general operations for workpiece in Figure 13-2.

its impact on the total variation that the final product exhibits, not an easy undertaking to accomplish as the number of DRFs multiplies.

The company's team was asked to establish parameters to describe the current situation and a set of goals for the project. Table 13-1 contains a list of the team's concerns. In particular, the major items of importance included the recognition of the family of parts and the approximately 40% of the processed material that did not conform to the documented specifications.

A more fundamental concern underlies this list. The drawings used to establish conformance were geometrically vague; they did not define the required rela-

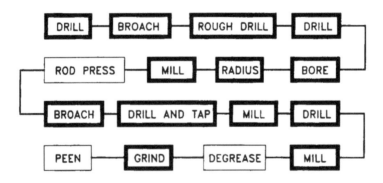

☐–INDICATES OPERATION REQUIRING NEW SETUP.

FIGURE 13-4   Schematic of manufacturing cell.

TABLE **13-1** List of Project Team Concerns

Approximately 40% of all production is rejected.
Drawings are geometrically incomplete.
Interrelated geometry cannot be assessed.
Uncontrolled and unidentified sources of variation.
Variation introduced by 13 individual setups.

tionships among the functional features. This led to reworking "rejected" parts where possible or using parts that could not be made to meet the specifications on the prints but, somehow, were still functional—all in all, not a situation that instills confidence in anyone's ability to describe the desired product.

An important piece of information from management's perspective was cost accounting information that quantified the cost of nonconformance. While the figures the team quoted probably did not include all the pertinent costs, sufficient dollar amounts were involved to attract attention and gain approval for the project. Expense dollars add up quickly when 40% of the product is initially rejected.

The goals listed in Table 13-2 are very concise. After finding that some 13 separate setups were involved in the processing of a part, it was decided that a single, universal fixture would be desirable. The design of this fixture would then allow the second goal to become reality, a single setup for all the operations. The most critical element in terms of achieving these goals was the reduction of the 56 different products to a single, parent component. This could then be modified to satisfy different customers, allowing major portions of the processing to be standardized, and yielding more consistent results than had been experienced.

13.3 COMPONENT DEFINITION

To begin the structured process, the first step involves describing the component. A clear and complete definition of the desired product is derived from an unambiguous geometric description. The necessary drawings to achieve this are developed by the concurrent engineering team using geometric dimensioning and tolerancing. As the team reaches a decision on the form the component should take,

TABLE **13-2** Goals Established
by the Project Team

One Universal fixture.
One setup for all operations.
One standard part.

this phase of the design process prevents alternate interpretations from arising later in the product cycle.

In Chapter 5, on component-level design, the presumption is that a design layout existed showing how the part related with its mating assembly. Since this project dealt with an existing product, completed assembly drawings and physical examples were available that allowed the functional interfaces between mating components to be determined. Equally important was existing information related to process capability that would be used to guide elements of the design.

The team's first consideration was the existing product drawings (simplified and abridged in Figure 13-2). These did not provide adequate description of the necessary geometry (its shape and size) or the allowable variation of the geometric features. The simultaneous engineering team dealt specifically with the lack of precision in defining the part's location in three-dimensional space. This encompassed two design deficiencies: the ambiguous communication of the part's interfaces that effected assembly (the functional DRF), and the ill-defined interrelationship of the individual features on the part.

Further highlights of the product and process review identified a number of implied datums that could be inferred from the drawing; none gave adequate definition to a single Cartesian coordinate system to be used in the design and fabrication processes. This returns to the comment that the component's location in space was not defined; there was no consistent method of kinematic control specified or used to produce the part. The high rework and scrap rates were the direct result.

A more pertinent concern arose at this point in the project. The rough material the company used to begin the process was a forging provided by an outside supplier. The lack of a single, explicit Cartesian reference frame in the product drawings was compounded by the introduction of a separate forging drawing with its own reference features (Figure 13-5). The forging supplier had never been questioned about the product definition and its impact on his process. As a result, he formed his own ideas about what was called for in the forging drawing. Unfortunately, these ideas did not coincide with those of the customer and thus introduced a major source of variability. No definitive method of qualifying the forgings existed, because none was specified on the drawing. As a consequence, personnel from the forging company were added to the concurrent engineering team to resolve these issues.

13.4   THE SIX-STEP PROCESS

At this point the six-step methodology (Chapter 5) was used to develop a precise and producible definition of the component. The original part detail is shown in Figure 13-2 in abbreviated form. The use of GD&T is notable by its absence— as it was on the original drawing. The most glaring error is the lack of a DRF

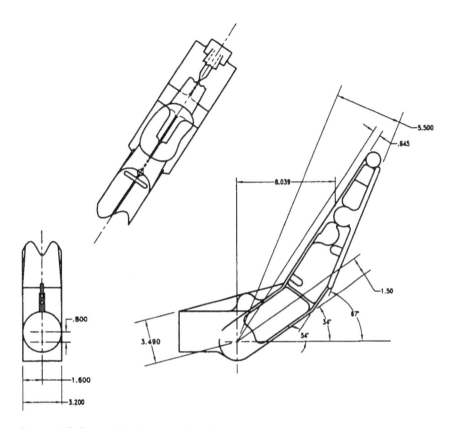

FIGURE 13-5   Abridged detail of forging.

to provide the definition's foundation. Further limiting any consensus on a single product definition is the number of implied datums. Neither the machined component nor the forging can be defined without clearly specifying the functional DRF. This was not done by the original design.

The ensuing discussions noted that the product and its process, in the form it then existed, used 13 different setups and fixtures. Both the selection and sequence of these operations never made it apparent that any of the geometry was to be interrelated. In a strict sense, these individual setups introduced 13 individual process DRFs. Thus, no two finished components were alike. There was so much process variation that the result was the rework and fitting-up expense experienced in the assembly operation. In many cases, this cost of nonconformance could consume up to 24 hours per assembly.

Confirmation of these problems was found in comments by the operators such as

Locates differently.

Locates differently yet.

Isn't easy to hold tolerance.

Most of our fixtures over here locate differently.

If holes aren't drilled straight in the part, when we locate off these holes
it throws that off alignment and we end up with oval holes.

If the forging is warped . . . can snap all four taps.

The last item is a critical comment because this particular operation vividly illustrated the use of a different setup for the primary datum. A number of holes were
drilled from one side, the part was moved to another setup, and then four of the
holes were tapped in a second station on the machine. While the tapping was
being done, the set of holes on another workpiece was drilled in the first station.
This technique allows both drilling and tapping operations to occur simultaneously. It does not, however, use the same setup for the drilling and tapping
operations. Processing problems (i.e., broken taps) result. This was especially
true in the case of a warped forging, which might break all four taps.

The interesting thing these comments show is the acknowledgment by production personnel that the problems encountered were caused by all the different
locating methods. It appears that the machine operators were well aware that they
were holding the parts every way but loose. It was just as obvious that they had
not reached the stage where this understanding could be translated into action to
resolve the problems.

The first task in front of the team was to establish the DRF for the part.
As mentioned, there were initially 56 different drawings that contained as many
as 20 implied datums. Many of the dimensions originated at points in space and
were used to locate a variety of features and angles. Consequently, there was no
unique product definition from which the company or its supplier could work.
Furthermore, even if consensus was achieved, the use of imaginary locating
points presents tooling design problems to manufacturing and inspection. There
were no real features equivalent to the imaginary locating points used in establishing desired part location.

After much discussion, the team decided on the product definition shown
in Figures 13-6 and 13-7. The first figure shows the specification for the forging.
The project team arrived at this forging definition after extended discussions.
They identified the need to refine the forging surface by

Adding to the forging definition datum A with flatness control

A 120° V locator to establish datums B1 and B2

A stop used to establish datum C

Creation of a universal qualification fixture for assessment of forging quality and for the fixturing of the forging during manufacture and measurement

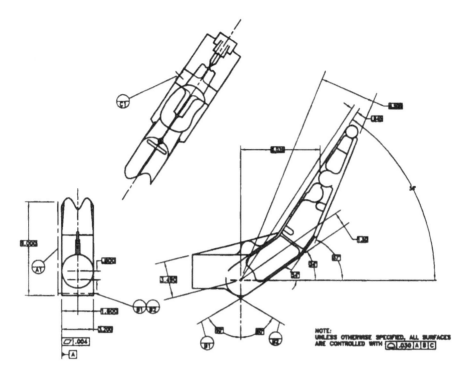

FIGURE 13-6 Abridged detail of forging with GD&T added.

The DRF that appears on the drawing finally enabled the forging supplier to measure its process capability. The decisions on what characteristics to gage for acceptance of the forged product were made by the forging company in consultation with the team. Because of the nature of the forged surface, the DRF is established using the datum targets mentioned above. These provide kinematic definition to the forging, give reproducible location to the component, and reduce measurement uncertainty to a level that allows the supplier to monitor and work with his process. Until this definition was provided, the use of different locating features by the company and its suppliers had introduced so much measurement uncertainty into the process that any acquired variables information was useless. The drawings provided absolutely no guidance for determining product conformance or process improvement and may have actually been misleading.

In looking at the detail of the machined component, it is apparent that the same reference frame used on the forging is also incorporated into the machined part definition. Where possible to do this, it eliminates the need to create a process DRF and, subsequently, to machine a separate functional DRF requiring a controlled relationship to the process DRF. When a process frame is used to get the

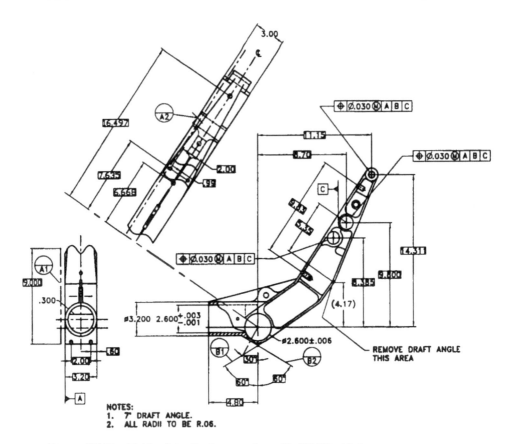

FIGURE 13-7   Abridged detail of support bar with GD&T added.

part out of the rough, it must be related by geometric controls to the functional frame. Each of the chained DRFs adds to process variation—as it obviously does to the cost. If any part of the DRF chain can be eliminated, then there is a reduction in the potential variation that can occur; there is also a corresponding increase in conformance and yield.

Once the DRF is established for the machined component, all other features that are part of the product definition can use it as a reference. The resulting product definition has been documented using a national—and international—standard. The definition's value results from the structured engineering approach that incorporates "good manufacturing practices" into both the specification of the product and its allied manufacturing processes. The elements forming GD&T contain the wisdom of years of manufacturing experience and lead designers through a standardized and, more importantly, logical design process.

At this point in the methodology, the DRFs have been defined for the forging and the end product. In this particular case, the end product, using the same reference frame as the forging, has what might be described as a process DRF. However, it can be demonstrated that this DRF, even though it does not include functional datum features, reduces manufacturing variability. This occurs by using a specifically chosen manufacturing DRF and then generating all the features that form the functional DRF and the remaining functional features in a single setup. Such a technique eliminates the effects of multiple setups and reduces errors to those inherent in the machine's capability. The result is a satisfactory relationship between the functional features and the actual functional DRF.

It should be mentioned that in conjunction with the universal fixture design, the team identified a number of discussion areas and action items to be addressed by the firm and the forging supplier. Clarification was needed on each of these items since they would ultimately affect the execution of the contract between the firms.

1. Clear definition of the purpose of the universal fixture
2. Acquisition and use of variables data.
   a. Determine dependent variables.
   b. Process capability measurement.
   c. What does forging company need or want to measure?
   d. How will these measurements be taken?
3. Calibration of the universal fixture
   a. Forging master used to calibrate indicators on the forging fixture
   b. Master machined part to ensure correct CMM setup and to compare production parts after they are machined
   c. Frequency of calibration
   d. Method of calibration
   e. Spare universal fixtures
4. Documentation
5. Purchase agreement
6. Fixture maintenance
7. Gaging/inspection policies
8. SPC charts required

The layouts for both the forging qualification fixture (Figure 13-8) and the machining fixture (Figure 13-9) result from the second step of the design process. Both fixtures utilize the same DRF called out on the details of the rough stock (the forging) and the finished workpieces. It is readily apparent that the critical information necessary to design the fixtures includes datum features identified on the finished part details. For both pieces of tooling, the primary datum is created by the use of tooling plate. Process characteristics related to the forging were the basis for deciding how to establish the primary datum. The forging

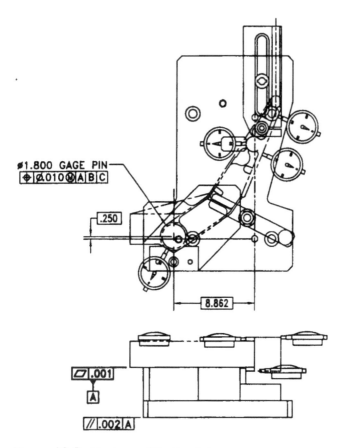

FIGURE 13-8   Forging qualification fixture.

method that was chosen controls this surface, which provides the datum feature on the part, for flatness within 0.004 in. when the feature was produced by the bottom side of the die. Functionally, this provided a primary surface of acceptable quality that could be utilized without further processing. The secondary datum is simulated by the 120° V-block with the tertiary feature generated by a stop. The angle of the V-block was again based on a functional requirement. This would allow the location of the forging to maintain a wall thickness of 0.100 in. These tooling elements mimic the theoretical datums shown on the component drawings. As pointed out earlier in the book, the component drawing contains the seeds of the tooling design. The discretion of the tool designer is limited to areas that will not have major, and possibly unappreciated, effects on total process variation and product functionality.

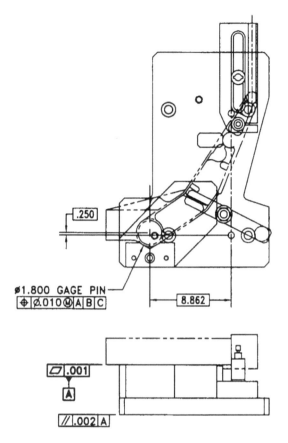

FIGURE 13-9 Machining process fixture.

Once the part is located within the fixture, the appropriate clamps, supports, and other elements necessary for good tooling design are added. In this particular example, the fixtures also included the means (i.e., a process-related DRF attached to their base) to locate the fixtures relative to the machine tool. The intention was to use the universal fixture in a manner similar to the way a pallet is used in a transfer line. Thus, in each of the fixtures, a DRF is established by the underside of the tooling plate (the primary datum) and two bushings that form the secondary and tertiary datums. These latter items mate with locating pins in each machine setup. Other elements of the tooling design are then related to this tooling DRF.

Once the clamps and supports have been added to the basic design, a stress analysis was performed on the part-fixture system to determine if the clamping arrangement created unacceptable deformation in the part, preventing the processed part from achieving the specified level of control.

The next task was to determine which of the critical and major characteristics of the part required inspection. For this component, it was decided to require variables data taken using a CMM. To increase the accuracy of the inspection information and to provide predictions relating to assembleability, the team mandated that an inspection fixture be used in the measurement process to reduce location errors. This turned out to be a straightforward activity since the fixture designed for forging qualification and manufacturing could continue to function as a universal fixture for measurement purposes. The only distinction between the earlier forms of the fixture and this inspection fixture would be the level (i.e., magnitude) of the controls that would be applied to reduce measurement uncertainty. It has to be made to more precise tolerances, but the same type of controls, to reduce this uncertainty.

An additional idea contained in the design of these fixtures is a workpiece blindside. One side of this component was intentionally designed without any features requiring machining. This allowed the blindside to contact the surface of the fixture, establishing the primary datum and eliminating the need to reorient—which would create another process DRF—the piece to do any machining on the primary surface. An additional advantage of this design approach results from its effect on engineering changes. Since the workpiece does not have any machined features on the locating surface, and with targets used to create the secondary and tertiary features, subsequent engineering changes will not require any changes to the universal fixture.

The result of all this activity is a tooling package that incorporates complete definition of the components that are made and the manufacturing and inspection tooling needed to support the design.

13.5 THE RESULTS

As already implied, this is one of those success stories that is almost too good to be true. The universal fixture satisfied all the team's goals. Major improvements in productivity were achieved in each step of the process and a number of key issues were resolved.

In the first instance, the team put together an unambiguous definition of the final product that specifically tied the description to the function of the part. Taking 56 different, yet similar, designs and incorporating their features into three part details, which yielded a single parent forging, reduced product proliferation. At the level of the forging, the clear description of the component (based on GD&T) and its subsequent use in the design of the qualification fixture achieved a 13% reduction in the cost of the forging. This resulted from smarter design procedures and not from new manufacturing processes. The emphasis is on "smart" design techniques that are more cost-effective than the more typical unstructured methods when costs are computed over the life of the product.

In the cell-based machining operations, the universal fixture (essentially the same as the qualification and gaging fixtures) reduced manning requirements by 50%, resulting in significant annual cost savings and cutting unit cost in half. This is even more impressive when considering that no new capital equipment was involved. Because the product had been in production for two decades, the cost of the new fixtures could be attributed to maintenance rather than to new capital investment since the existing fixtures would require eventual replacement.

Looking at the flows through the cell, the universal fixture reduced process cycle time by 50%, from approximately 30 min to 15 min. Even more impressive was the cycle time of 8 min on a 3-axis CNC machining center that could replace the cell once the universal fixture was available.

The results included the part design, manufacturing methods, and verification techniques all incorporated within a unique system of product development. Such a system is effective only within the context of a concurrent engineering team. Without the team effort, it is impossible to get the communication needed to master the development process. From a purely technical standpoint, the geometric description of the part using internationally accepted standards provides that unique set of information that ties the part design to the design of the manufacturing and verification systems. The consequences are a striking example of the use of intellectual resources rather than physical assets to achieve large productivity gains.

.

14

Implementation and Process Improvement

14.1 INTRODUCTION

With the foundation and structure of the integrated techniques in place, the final task is to show how these structured methods may be implemented and used to support an integrated product development process. Placing the concurrent engineering methodologies within such a framework serves to reinforce their power and provide an economic argument for their use.

This text presents a set of powerful methods that take the technological activities comprising the product design process and integrate them with the firm's human resources. The primary goal of these techniques is to develop the company's personnel, overtly linking the firm's unique technical knowledge with its product development organization. Even the most successful management techniques do not provide the depth of structure needed to accomplish this. The combined use of concurrent engineering based on geometric control and team-based continuous improvement provides that structure. For an organization to prosper in a competitive global environment, these methods provide the details—the numbers—necessary to give definition to the product and allied processes at the time and level appropriate to the firm's economic goals.

The material focuses on one major technical theme, the use of the product definition to drive integrated product development programs. The discussion now turns to the importance of this theme in managing such programs and then examines how the techniques can be introduced to an organization.

14.2 WHY FOCUS ON THE DEFINITION?

The previous chapters show how to create a complete and producible product definition. From a management viewpoint, how this information drives the design and improvement processes must now be demonstrated. The rationale for emphasizing the definition falls into two major categories: the human resource or organizational area and the technical area. Of these two, the organizational concerns will ultimately have the greatest positive impact on the company's success.

14.2.1 Management Tool

The primary reason justifying use of these integrated techniques is that they lend structure to the initial design of the product, which can then be extended to subsequent design and improvement efforts. A common thread is woven throughout the product life cycle, achieving the continuity lacking in the project management of many products. These techniques are rationalized management tools that provide a structured product definition and organize the necessary baseline information utilized in both the primary design efforts and downstream processes.

Two overarching themes dominate use of these techniques. First is the integration of the "design and build" portions of the product cycle to eliminate inefficiencies introduced by unproducible designs. The unintended costs that manifest from these inefficiencies occur in the traditional sequence where the product is designed in isolation and "thrown over the wall" to manufacturing. Traditional design, with its clear demarcation between the design and manufacturing functions, is a tremendously inefficient way for the organization to operate and introduces many organizational problems. The second theme involves the desire to move product design and its allied processes away from a creative art to something more structured and scientific in nature. The more rational the design cycle, the more predictable the results.

14.2.2 Communication

Although already mentioned, the use of the Y14.5M standard as a common language cannot be overemphasized. Because design, manufacturing, and quality organizations use it across many industries, the standard provides a medium in which concerns and decisions relating to the project can be communicated.

A more important point needs to be made. This common language is intricately bound to the design and improvement methodologies developed earlier. The correct application of these methods requires the precision of the geometric control language. However, to correctly use this language, the suggested design methodologies must be incorporated into the firm's procedures. In essence, the two cannot be separated: One is linked to the other by necessity.

As an element of the continuous-improvement process, the geometric language ensures (1) that the project teams can identify candidates—in the initial and later stages of the product life cycle—for improvement and (2) that these efforts are not affected by the use of fuzzy language. The hazy thought processes that follow from imprecise communication of the product definition reduce the opportunities that can be identified and restrict the results that occur.

14.2.3   Education

One of the more long-lasting organizational benefits is educational. With the rapid changes that take place in the general business environment, training and development of technical and manufacturing personnel become major undertakings. While formal training sessions can impart general principles and concepts, much of the information and experience that are unique to a given company are either difficult or impossible to impart in this fashion. This information may well be the final arbiter in determining the market success of the product and of the company.

The concurrent engineering team is an excellent mechanism to deliver the specific training that should be provided as an ongoing investment in personnel. Overlaying structured design methods—based on geometric control of actual projects—with team activities allow the individual team members to be educated using sufficiently complex situations to bring them to optimum design efficiency. This yields immediate economic benefits provided by work on real projects and is more effective than simplistic classroom exercises.

A further benefit is the body of proprietary knowledge developed by the technical activities the team undertakes. Methods, procedures, and informal organizational links are created that help bring the project to a successful conclusion. These results, if attended to properly, are not transient but long-term organizational assets. In fact, it is not inconceivable that an accountant can place a value on these assets and include them in the company's balance sheet. The leverage obtained through this type of product development may well be one of the major factors in justifying these methods.

14.2.4   Problem Solving

The concurrent engineering process requires enhanced problem-solving abilities. The most visible indication of these abilities is problem solving by a team rather than by an individual. Without the team's common and complete understanding of the physical reality of the product, much of the subsequent development activity lacks direction. In essence, the development cycle becomes one large experiment where results are obtained by trial-and-error, an approach that is no longer economically viable in a time of increased competition.

More to the point, informal and unstructured development procedures are not the hallmark of outstanding management. The specific types of information and decisions related to the product and process definition require recognition of many of the problems that classical techniques have left undefined—those troublesome numbers again. These problems are left to be solved in the course of full production. The improvement process coupled with concurrent design forces early recognition of these concerns at points in time and at levels in the organization that can provide acceptable solutions. Production is not disrupted by the need to sort out problems that occur due to incomplete product definition.

The integration techniques provide the means to both initiate and control problem solving. The structured approach can be used as an entry point into the early stages of design and process improvement by formalizing general concerns and reducing them to quantifiable concepts. As pointed out earlier, variability is the most important of these concerns. The techniques provide a well-ordered set of management controls and design procedures that place reasonable boundaries around the areas of variation considered as candidates for optimizing the design. They are also the mechanisms that lead to specific improvement projects. The underlying logic of the design techniques provides the means to control the project.

14.2.5   A Benchmark

Continuous improvement requires recognition of specific business processes and leads to identification of opportunities for process improvement. In giving a physical reality to a product, the product definition—which, in its broadest sense, includes information about both product and process—provides a reference from which to begin the improvement process. This statement encompasses a much broader view of the improvement process, seeing elements of it even in the early stages of design.

Actual production is usually the result of compromises associated with practices that have evolved over the years, yielding a less-than-optimum design for the production process. The design layout, containing all the geometric information provided by the dimensioning and tolerancing techniques, creates a benchmark from which elements of an idealized manufacturing process can be defined. As a consequence, a clear definition is provided for the various elements that might be contained in an optimum process. This provides a target for the concurrent engineering team as well as a basis for formulating a continuous-improvement program.

While benchmarking in current usage implies finding the best product or process available to use as the standard, the product/process definition that results from the concurrent engineering activities allows the product to be self-referencing. Thus, many of the problems that occur in trying to benchmark a product where no process-related information is available are eliminated. Benchmarking of the definition's process elements is driven by the idealized process sequence embedded in the geometric design. The problems encountered in trying to get externally generated information based on what is probably proprietary knowledge are overcome. In the event that outside sources of benchmarking information are available, these can be folded into the improvement process while still using the internally generated product definition.

14.3 STAGES OF IMPLEMENTATION

Once an understanding is acquired about how a structured process may be justified, the next step is implementation. What follows assumes that a traditional design structure is already in place. Whereas the structured techniques are as readily applied to the improvement of existing products as they are to the design of new products, existing product development systems will provide the most challenging test of their use. To meet the challenge, a core group of individuals must embrace the integrated process and become its advocates. Without their fervor, the techniques will likely become a discarded fad.

14.3.1 Audit Existing Design Process

Whether attempting to apply the techniques to a new or existing product, the starting point is to audit current design practices and determine how and where the important elements of the design are being specified. At the geometric level, this audit can be performed using the notion of design refinement discussed ahead.

As with many other fields, the long-term trend has been to have engineers concentrate on areas of specialization; many times, support and manufacturing groups assume responsibilities that should belong to the engineers and designers. As the engineers and allied designers have become more involved with analytical design, much of the information required for a successful product is provided by an informal design organization. This informal organization includes elements of purchasing, manufacturing, quality assurance, suppliers, and other groups. In most cases, it is both unstructured and unrecognized. Consequently, the typical product definition found in the engineering drawings and databases is not sufficient for successful completion of the project. It has to be fleshed out as the design moves downstream, usually done in an uncontrolled fashion.

TABLE 14-1   Levels of Tolerance
Refinement
_____

Size
Limits of size (Taylor's principle)
Datum reference frame
Geometric controls
   Form
   Orientation
   Location
   Profile
Surface texture
   Waviness
   Roughness (systematic)
   Roughness (random)
Surface integrity
   Crystalline structure
   Lattice structure
_____

Source: Adopted from DIN 4760.

The concept of geometric refinement can be used to structure the design process audit. Such an audit would look at the point in the product's life cycle where the various elements that serve to refine the product definition are added. The audit could be organized on the basis of form deviation contained in the German standard, DIN 4760. The levels of deviation (variation) shown in the accompanying table (Table 14-1) have been modified for use here; as a basis for an audit, they provide some idea of who creates this information, when it is created, and where it is documented. From the standpoint of the economic success, these are all critical questions.

The purpose of the audit is to determine whether a complete product definition is assembled and documented at any time in the cycle and, if so, how this occurs. If the definition is assembled in piecemeal fashion and the information dispersed throughout the organization, economic inefficiency is built into the process. This should be addressed through more organized design procedures. The audit proceeds from the level of shape and size information and continues through the levels shown in Table 14-1, which provide three-dimensional control. Certain products require refinement to the material structure, but the majority of products can achieve significant benefits by proceeding only to the surface texture level.

The audit has three ultimate goals:

1.   There is an attempt to learn if design intent—functional requirements—is carried throughout the process. With dispersed or segregated decision making, this may not always be the case.

2. Producibility is assessed by ascertaining if the realities of the production processes are recognized in achieving product function. The complete absence of geometric controls or, of equal likelihood, a request for unrealistic features or levels of control discredit the integrity of the product definition. The latter leads to subversion of the design intent, with manufacturing personnel ignoring the more refined and difficult-to-achieve specifications.

3. The audits give quick readings on whether the product definition takes physical realities into account by anticipating effects of individual tolerance interaction on total product variation (an uncertainty budget).

14.3.2 Education in Documentation Principles

Should the audit reveal that the design definition is put in place over time by isolated individuals and functional areas, then the first step in implementation would be to provide an overview of product documentation techniques (see Table 14-2). The logical starting point for this would be the concepts contained in and allied with the Y14 graphics standards. Additional areas of exposure should include the B4 standards (now inactive) on limits and fits, the B89 standards that deal with applied dimensional metrology, and the B46 standard on surface texture. Together these standards provide some of the theoretical basis for structured design techniques and are well-established methods that should be incorporated in any professional design process.

As the move to integrated systems—computer-based or not—occurs, it is also important that the design organization institute some form of configuration management. Along with the obvious operational benefits, this serves to emphasize the interdependence of the product information and is used to create discipline in the product design cycle. Team-building activities should be introduced at a later stage after the structured design techniques are implemented. The design methods can easily be instituted as they build on existing activities that are the individual designer's primary responsibility. Team responsibilities will require more radical behavioral changes. Delaying their imposition until after the intro-

TABLE **14-2** Selected Domestic Standards Applicable to Structured
Product Documentation

B4	Standards	Limits and fits
Y14	Standards	Engineering graphic standards Y14.5M, Y14.8M, and Y14.36
B46	Standard	Surface texture
B89	Standards	Applied dimensional metrology B89.1.12M, B89.3.1, B89.6.2, B89.7.2

Note: There are specific ISO series of standards that are equivalent to the above examples.

duction of structured design may better foster acceptance of these behavioral changes.

14.3.3 Senior Management's Support

The preceding stages are readily justified and can withstand scrutiny. The suggested tools should be found in *any* engineering design effort, no matter what level of sophistication it has. Without standardized documentation and communication, much of the proprietary knowledge that results from putting a product into production is lost; as a result, the system keeps reinventing the wheel. Thus, management's agreement to take these first steps can be easily acquired based on the activities conforming to accepted engineering practice.

The difficulty in getting management's approval to start the program's more collaborative elements is likely to arise as a move is made through the levels of refinement beyond the shape and size information found on all engineering drawings. As the information density is increased (i.e., the design becomes more refined), the apparent costs incurred in initiating the design process also increase. It naturally takes longer to provide the additional information necessary to refine the product definition when compared to traditional techniques. It also requires more highly trained engineers and designers. Furthermore, because this information is based on part function, it needs considerable more thought and calculation to arrive at the desired controls. These activities take time, inflating the cost of early project stages.

Senior management must be convinced that the expenditure of additional resources at the front end of the cycle will have economic benefits that reduce total project costs. One suggestion to garner management's support is to show that the information needed for the initial product definition is generated within all existing projects at some point in the cycle. The earlier that this information is incorporated into the definition and documented, the less it costs to make changes. Some dated estimates of engineering changes run from $3,600 per change in the electronics industry to $14,000 in the aircraft and aerospace industry (Vesey, 1991). When these estimates are adjusted to current dollars, it should be easy to convince managers that the trial-and-error process of product design is economically inefficient at best and suicidal in an extreme case.

With more highly skilled workers required to complete the product definition early in the design cycle, knowledge these individuals accumulate through education and experience becomes one of the more important company assets. The goal of management is to get access to this wealth of knowledge in a manner and at a time appropriate to the needs of the development cycle. Senior management must be convinced that the most efficient way to tap into this source of technical and organizational knowledge is through the use of concurrent engineering concepts based on geometric control principles.

14.3.4 Require a Structured Design

The design process should take advantage of structured techniques that evolve when geometric control is applied. Having achieved top management's support for a design process incorporating use of geometric control and engineering teams, the structured baseline design must be implemented and enforced.

Management must insist that the assembly's design layout be completely defined before specific components are designed. The complete conceptual definition of the product must be reduced to an engineering design (the design layout) with sufficient detail to define and analyze the derivative component designs. The design layout provides the baseline information from which all further product decisions are derived.

Previous design approaches—the reality, not the ideal—relied on the simultaneous design of individual components and the assembly of which they were members. The typical argument justifying this approach was that it expedited the design process and moved the product into production more rapidly. With the increasing emphasis on time as a competitive weapon, the pressure to design components prior to complete definition of the assembly is intensified. Unfortunately, what is really expedited is the production of faulty designs due to a lack of a complete product definition.

14.3.5 Core Implementation Group

The concurrent engineering team must be composed of the appropriate members to properly define the product. Individual members must cover the entire range of interests and functional areas that have impacts on the product cycle. They must also have complete authority to represent their constituencies. The ideal project structure has all the functions incorporated under a project manager who has direct lines of authority to these individuals such that they are bound to the project's success. An example of this type of management structure is found in the Japanese automotive firms and has led to quite successful product management outcomes (Clark and Fujimoto, 1991).

In any event, whether it is a matrix structure or a cohesive and formal team, the group must contain members from all critical organizational elements. Representatives of each discipline must participate on the product development team.

14.3.6 Training Issues

The next important undertaking is the creation of common ground from which to approach the design work. As explained earlier, the concepts contained in the Y14.5M standard provide a language in which to conduct discussions of project merit. This is a precise language with enough complexity to require formal training. All participants must have a common level of facility with the language. The main issues involved in this training are covered in depth in earlier chapters. The major themes include the use of datum reference frames, the general concept of geometric control and refinement, and verification methods.

14.3.7 Identification of an Advocate

There are a tremendous number of techniques—magic bullets—vying for management attention. If these ideas are forced on an organization rather than being generated from within, they fall on infertile ground. Furthermore, where someone is chosen at random to implement the methods, there is little possibility that anything positive will come of the effort. The cultivation of these ideas and the resources that bring them to fruition need careful attention that can only be provided by a "believer."

To gain acceptance, the structured techniques must be carefully integrated with the existing product design cycle. The following material provides some suggestions related to what is arguably the most crucial decision necessary to reap the ultimate economic rewards.

Choose an Advocate. As the use of these methods is not the norm in industry, the rationale behind applying them may not be obvious. In extreme cases, there may be individuals with 30 years or more of traditional design experience—probably with considerable success—who see the use of these "new" concepts as unnecessary. The inertia that prevents implementing these changes can be overpowering. If an edict comes from upper-level management forcing change, people will "go through the motions" for some period of time until either management becomes frustrated and gives up or sufficient negative experience accrues that proves the method should be abandoned.

At this juncture, management must realize that it is seeking behavioral change in addition to design technique change. More pertinent, these changes will not occur by fiat. They must be fostered by example; to set such an example, an advocate must be appointed.

The selection of the advocate is an extremely important and, quite likely, difficult decision. The chosen individual must be a highly disciplined designer. The candidate must already have an operable knowledge of the structured techniques. No benefits will result if a technical novice is asked to institute the change in design procedures.

On another plane, the chosen advocate must have good interpersonal skills. These are necessary to interact with senior designers who have to be convinced that what they have been doing for all their professional life is no longer acceptable under the new competitive rules. The advocate must genuinely enjoy helping people learn if this change is to succeed.

Provide Advanced Training. Training opportunities must be made available to the advocate in both advanced applications of geometric control and group dynamics. Enough has been said about GD&T to see where this fits into the implementation phase. More should be said about group dynamics.

The team is being assembled to deal with the complete product development cycle. The ultimate success of the development process depends heavily on team cohesiveness. To foster cohesion, the advocate must have a working knowledge of group dynamics and the necessary communication skills. In a small firm, this person may well be both project manager and facilitator. A larger organization may have a sufficiently large number of product teams where the advocate functions solely as a facilitator. In this latter situation, where the authority of the project manager title is missing, many of the interpersonal skills needed to accomplish the team's goals will probably be based on inherent ability but may also be acquired and enhanced through formal training.

Speaking to the training needs in the technical area, a certification procedure (ASME Y14.5.2) covering knowledge of geometric control has been established. The procedure provides for different levels of certification. Described liberally here, the first level involves the ability to interpret the controls placed in the product definition. After acquiring such knowledge, the individual would move to the next level, which involves applying the controls to a design. A final step would be the ability to teach the material, a trainer's level, although this is not currently part of the standard. The advocate should be someone with application ability and experience who has the desire to reach the trainer's level. It would be impossible to facilitate the team's discussions and negotiations without the requisite knowledge. The prime mover in these discussions is the technical information necessary to provide a geometric product description.

14.3.8 Management Support

From what appears so far, it should be obvious that active management support is crucial to begin the process. Timing changes within the expense stream now move costs to the initial stages of design, making management's tolerance of this front-loading of expenses imperative. Without tangible, comprehensive management support, the techniques should not be implemented. This support requires a champion on the senior staff: Without this champion, program implementation takes the form of an edict that will not foster the desired behavioral and technical changes.

Management must put mechanisms in place to indicate the high level of managerial support given to these techniques and also to make the techniques highly visible. One of the most obvious mechanisms is to place the advocate in a separate position reporting to upper management. The logical situation would then be to vest the advocate (possibly the project manager or a senior member of the project team) with the authority to require the changes in design technique from which behavioral changes are expected to follow. The resources devoted to filling such a position, and the management access the position has, would demonstrate the high level of commitment necessary for success.

While less visible, other ways of supporting the effort include senior management's participation in design reviews and audits, insisting that a complete product definition be developed early in the design cycle. In the improvement area, the use of the completed definition as the benchmark against which changes may be analyzed and judged will further reinforce the need for structured design.

A more formalized way of supporting implementation is to require vendors to implement procedures that parallel in-house activities, becoming versed in the techniques of geometric design and continuous improvement. This can be done most readily by inserting GD&T requirements in vendor audits and, ultimately, as terms in purchasing contracts. Much of this is already in line with the philosophy found in many national and international standards. ISO 9000 is a good example of a standard-driven impetus to foster these concepts.

14.3.9 Controlled Implementation

In many environments success comes more readily if changes are introduced in a rational and logical sequence. The greatest probability of success in fostering structured design occurs when it is introduced in a controlled manner. As pointed out, this operates at two levels.

On the technical level, it would be rare for any but the largest organizations to have more than a few individuals with extensive facility in the use of geometric controls. This group needs to be expanded. Because of the method's power, the underlying language is somewhat complicated and requires effort for mastery. Traditional product design practice, achieving robustness in the product through large tolerances and allowances, does not encourage such learning. Of even more importance, management's acceptance of the results of the traditional process—which no longer yields competitive quality levels—convinces designers that their approaches are acceptable and need not be changed. The economic reality now requires that this stance be reassessed. However, the underlying problems are not sufficiently recognized to bring about wholesale organizational change; many product designers do not understand the limitations of their methods and are not ready to take up the challenge.

At the organizational level, department structures work against wholesale introduction of these techniques. The appraisal methods that complement such structures help feed parochial interests and make it difficult for consensus to evolve. Without consensus, the team's ability to get things done is impaired.

To overcome problems in these two arenas, a project of limited extent should be used to introduce the structured methods. One suggestion is to overlay the methods on a continuous-improvement project where an existing product offers opportunities for productivity and quality improvements. Preferably, the product should be one of low assembly complexity or a subassembly. The rationale supporting this suggestion deals with the likely alteration of production methods. This might include incurring additional tooling charges to make changes in the processes. The small scale of the prototype project places a boundary around the expenses that may be incurred in implementing the improvements and the time required to achieve results.

A further reason to keep the project scale small relates to the need to achieve some visible and quantifiable results before questions arise about the project's efficacy. Even in the improvement mode, the front-loading of expenses may make managerial advocates of the process question their support of the test. A manageable project scope reduces the time frame and, of more importance, may allow the team to predict the outcome of the project with a higher level of confidence. The initial implementation of these ideas must be successful. Even marginal levels of success may cause re-evaluation and possible cancellation of the attempt. Choose the first battles carefully. This is to be a quiet revolution that avoids the major risks usually associated with more radical change.

14.3.10 Upgrade Metrology/Inspection Capabilities

This book emphasizes the need to base the improvement cycle on a quantifiable definition. This presumes that once the necessary geometric design is provided, some means to verify conformance exists. Without high levels of accuracy in the ability to verify the controls (i.e., to measure things), manufacturing interchangeable product is more difficult. Less precise measurement processes increase the required level of production skills. In the extreme case, lack of accuracy moves manufacturing back to the craft methods of producing products: An individual craftsman produces an individual product through handfitting of each component. No interchangeability exists in this situation. Elements of this occur in the instance where a product is inadequately specified and the manufacturing people must provide fixes—more definition—in order to bring the product into existence.

It is also important to remember that a firm's metrology capabilities are an integral and necessary part of developing the product definition. The introduction of very sophisticated scanning capabilities now make it feasible to gather a great number of points in a relatively quick fashion. This high data density allows form errors to be assessed using CMMs and this capability to be included as a standard element in developing the product definition, including the companion manufacturing processes.

A key realization is that the measurement process is as complicated an undertaking as the rest of the manufacturing sequence and requires the same amount of attention and planning. Many production routings show the inspection process identified as a single step in the overall scheme with little more than the one-word description "inspection" and a gage number called out. Under such operating procedures, the inspection operation appears to carry as much weight as packaging the finished goods and moving them to storage.

As the general level of precision is improved and this enhanced ability is incorporated into the inspection sequence, the inspectors must not be left the discretion to randomly modify the measurement process. If the individual inspector redefines the process each time it is implemented or there is variation in the techniques applied across different inspectors, then the advantage of a more precise product definition is lost. Different answers are given for the same question and the wrong things are being verified. The method of inspection must be directly linked to the product definition.

An important change in the thought process occurs when the designers realize that the measurement process is subject to variation just as the more traditionally defined manufacturing operations are. Once statistical variation of the measurement process is accepted, then it becomes necessary to carefully specify the measurement setup. It is imperative to realize that a significant portion of the allowable product tolerance can be used up by measurement uncertainty introduced by variability in setup and inspection methods.

14.3.11 Review and Critique

After the first project is completed, time should be taken to objectively review what has been accomplished. This review is an application of the third stage of the plan–do–check–act (Shewhart) cycle. Because this was a project of limited scope used to prove feasibility, no surprises should arise during the assessment. However, because each organization is unique, much can be gleamed from the improvement project to guide full implementation of the methods and procedures.

At the organizational level, the assessment will be used to justify continuing implementation. This is a business decision that will require expressing the project's results in monetary terms. Because many of the economic benefits that accrue due to these techniques are measured in "dollars not spent," activity-based accounting procedures may need to be developed simultaneously with concurrent engineering. It would be an unusual firm where the executives rely solely on qualitative arguments to expend resources, allow training, and organize for full implementation.

Audits of the results should be performed covering both the technical and the human resource aspects of the prototype process. In the technical area, the audit should focus on actual product and process documentation. The auditors can readily see whether or not the geometric controls driving the design are being correctly provided as integral elements of the definition. The auditors should note the source of the controls and ascertain the stage of the design cycle when the geometric controls are determined. A more subtle look at the definition would determine how the actual process description deviates from the idealized design in the engineering database. Specifically, in a perfect world, a single DRF would be used in describing and fabricating each component. If this is not true for components selected for the audit, there should be documented justification for the departures and analysis of the subsequent effects expected in the downstream processes.

Another method used to audit the product's technical specification involves reverse engineering of the component that mates with a particular part. A significant amount of information concerning the functional characteristics of the mating component is necessary to describe any part. By assembling the information from the sampled product definition, the auditor should be able to describe many of the mate's functional design elements. If this can be done, then the original product definition was done adequately. If the mating part cannot be described functionally and, to some degree, geometrically, then the original definition does not contain a sufficient description to produce the workpiece. This technique can also be used to identify subjects for continuous-improvement efforts.

In the human resource area, the audit should focus on whether issues of training and placement were successfully resolved. Members of the concurrent engineering team are key players in the success or failure of the project. The auditors should determine if the appropriate people were placed on these teams and if these individuals were provided with the necessary resources, including training, to accomplish their assigned tasks. Not all individuals are able or willing to subordinate years of parochial interests to the needs of the product team. If such individuals are present, this should be recognized and alternate members brought on-board.

The final area to be audited is project management. Two distinct concerns are involved. First, there is no definitive way the project must be managed. Consultants and the literature suggest a variety of methods, ranging from the traditional functional structure, to matrix management, to a separate project team containing all necessary functional personnel. The object of this audit is to determine if the specific structure applied to the prototype project has worked. The guidelines used in the audit should not reflect what purports to be current management tenets but should be based on project outcomes. The audit team should identify the functional results and forward these for use in the decision-making process.

Second, the team leader and the advocate should be given performance reviews. Assuming that these individuals were selected on the basis of established technical competence, the review should be directed toward their success in coordinating the team's activities and team interactions with other groups. Assessment items should include the leader's ability to foster active involvement by team members. The review should also focus on the training opportunities presented by the project cycle. This last concern is important because the team members are learning problem-solving skills that will be generalized for application to other development projects. Training success is demonstrated by behavioral changes in the team members.

The results of the review should be shared. In particular, any positive aspects of the work—the successes—should be publicized. Also, on the basis of the results, any obstacles to the success of the techniques that involve planning, budgeting, or appraisal should be redesigned to support the product development system.

14.3.12   Expand Training

After the audit is successfully completed, the next stage is to expand the use of concurrent engineering and continuous improvement. To extend use beyond the prototype group, training must be provided in applying the technical tools that support concurrent engineering and also in team building to facilitate the continuous-improvement process.

Personnel should be trained in the use of the geometric control techniques in two phases. The first involves a basic introduction to geometric control, allowing the individuals to interpret existing product definitions. While the degree of application is at a low level, this is still a complicated undertaking involving significant interactions among the geometric language, the underlying design concepts, and the manufacturing and verification processes. This introduction must make designers sensitive to the interdependence of all stages of the product cycle before an adequate understanding can be developed. The second training level provides necessary experience in using these concepts to give definition to a specific product. The standard does not provide detailed guidelines that can be directly extracted from the printed document and thrown on the drawing. Each application is unique and requires a degree of intellectual effort—thinking—that cannot be avoided. While it would be nice to buy a set of specific instructions along with the standard, the designers must develop their own application guidelines. Thus, the second level of training is really an ongoing effort that continues as long as the firm develops new applications or products.

This last point contains additional importance for continuing the application of structured design and improvement. The training efforts must take place within the context of actual projects. Little of value will be learned by formal training utilizing artificial examples not relevant to a company's product line. The nature of the techniques requires learning-by-doing since the design process must be tailored to each product. The methods cannot be taught solely in the classroom but must take place in the context of productive design projects. Concrete examples of the company's products are needed to create both the specific techniques and supporting organization for successful product development.

A core group of trained individuals will be needed to expand the use of the methodologies beyond the experimental stage. Each individual must be qualified at the application level of integrated design so that he or she may serve as a facilitator as additional teams are created. Introducing others to the techniques can be accomplished by incorporating new individuals in the development cycle and the supporting concurrent teams. The new members can quickly begin to contribute in a positive way without having complete knowledge of the methods being used. This is an important point: The more rapidly a person can see the benefits of the techniques, the more likely it is that he or she will commit to the training necessary for mastery.

Management's desired outcome from the exercise (prototype and expansion phase) should be a host of creative problem-solving skills based on integrated design methods that can be applied to the development of a large variety of products. This cannot be achieved by management fiat. It can only be accomplished by the individual team members achieving personal goals.

14.3.13 Require Use

Under the guise of further training, the use of the integrated design concepts can be expanded to include additional projects and incorporated within standard procedures that support continuous-improvement techniques. This leads to the required use of these techniques in every stage of the product design cycle.

The continuous-improvement process can be used to foster companywide use of these techniques. As demonstrated, many of the goals and behavioral changes necessary for successful implementation of continuous improvement are directly fostered by integrated design. The emphasis on the product definition conforms to the Shewhart cycle referred to earlier. The base on which further improvement is focused is the product definition. What is now required is management's continuing motivation to keep these concepts and techniques alive. Constant attention to the activities and results of the process is needed to embed these techniques within the firm's standardized procedures of product development.

14.4 CONCLUSION

Most products take geometric form as the design is translated from concept to physical reality. The functional elements of geometry must be provided by the design team in a timely fashion if production is to proceed in an efficient and economic manner. The techniques described in this book allow this to happen in a well-managed way. Additionally, key causes of product variation are identified, allowing prediction of the effects of variation on product function. This leads to design alterations where necessary; or, at a minimum, knowledge of this variation will quantify elements of risk in the design so that the appropriate business decisions can be made.

The product definition provides a set of numbers (i.e., metrics) that allow technical solutions to be found based on objective methods rather than visceral responses. "Getting to the numbers" requires both a team structure and a common language with which team members communicate.

The design language is necessary from an engineering standpoint to provide (1) the fundamental information (dimensions) required for the part to take physical form and (2) knowledge of the dimensional variation that can be tolerated. The language and its companion methodology tie the design and manufacturing processes together, formally eliminating the artificial barriers that segregated these functional areas in the past.

The design methodologies work best in the context of teams. In fact, it is difficult to enforce design integration where concurrent engineering teams do not exist. The formal extension of the product definition and integrated design into the area of continuous improvement is the logical next step since improvement techniques require tight communication between large numbers of individuals and functional areas. If not solely responsible for making this communication possible, concurrent design teams greatly enhance its effectiveness.

The product definition provides a benchmark for the continuous-improvement process. This book's concurrent design philosophy is based on a foundation similar to that of the improvement process, although incorporated earlier in the product cycle. Hence, continuous improvement is effectively built into every phase. The emphasis on teams to create the product definition and to facilitate the improvement process provides the necessary organizational underpinning.

If critical product information is identified and documented early in the design cycle, the greatest economic benefits are achieved; if generated by trial-and-error during later stages of the cycle, the price is dear in lost market opportunities and avoidable expenses. The extreme and most unacceptable case is where these critical elements of the product definition are not fully described. The resulting variability of the product could well lead to its failure in the marketplace.

In summary, this book provides challenges and opportunities for creating intelligent manufacturing strategies. The cornerstone that underlies the integrated techniques is as follows: All the information contained in the product definition wending its way through the development process will eventually be generated for a successful product. Effective product management structures and controls the process to capture all the value added in each step of development. The correct use of intellectual resources—working smart through design integration—can supplant capital investment and reap maximum rewards for a successful product design.

REFERENCES

Clark, K. B. and Fujimoto, T., Product Development Performance: Strategy, Organization, and Management in the World Auto Industry, Cambridge, MA: Harvard Business School Press, 1991.

Vesey, J. T., The new competitors: Thinking in terms of "Speed-to-Market," Manufacturing Systems, **20**, June 1991.

Index